Progress in Inflammation Research

Forthcoming titles:

The Resolution of Inflammation, A.G. Rossi, D.A. Sawatzky (Editors), 2007
Endothelial Dysfunction and Inflammation, A. Graham (Editors), 2007
Adhesion Molecules: Function and Inhibition, K. Ley (Editor), 2007
Angiogenesis in Inflammation: Mechanisms and Clinical Correlates, M.P. Seed (Editor), 2007
The Immune Synapse as a Novel Target for Therapy, L. Graca (Editor), 2008
New Therapeutic Targets in Rheumatoid Arthritis, P.P. Tak (Editor), 2008

(Already published titles see last page.)

Chemokine Biology – Basic Research and Clinical Application
Volume II: Pathophysiology of Chemokines

Kuldeep Neote
Gordon L. Letts
Bernhard Moser

Editors

Birkhäuser
Basel · Boston · Berlin

Editors

Kuldeep Neote
Cancer Inflammation and Cell Survival
Eli Lilly and Company
Lilly Corporate Center
Indianapolis, IN 46285
USA

Gordon L. Letts
NitroMed, Inc.
12 Oak Park Drive
Bedford, MA 01730
USA

Bernhard Moser
University of Bern
Theodor Koch Institute
Freiestrasse 1
3012 Bern
Switzerland

A CIP catalogue record for this book is available from the Library of Congress, Washington D.C., USA

Bibliographic information published by Die Deutsche Bibliothek
Die Deutsche Bibliothek lists this publication in the Deutsche Nationalbibliografie;
detailed bibliographic data is available in the internet at http://dnb.ddb.de

ISBN: 978-3-7643-7195-1 Birkhäuser Verlag AG, Basel – Boston – Berlin

Part of Springer Science+Business Media
Printed on acid-free paper produced from chlorine-free pulp. TCF ∞
Cover design: Markus Etterich, Basel
Cover illustration: see page 7. With friendly permission of Steven L. Kunkel. Art work: Robin Kunkel, Department of Pathology, University of Michigan Medical School.
Printed in Germany
ISBN-10: 3-7643-7195-1
ISBN-13: 978-3-7643-7195-1

e-ISBN-10: 3-7643-7437-3
e-ISBN-13: 978-3-7643-7437-2

9 8 7 6 5 4 3 2 1

www.birkhauser.ch

Contents

List of contributors

James Broach, Princeton University, Princeton, NJ, USA

Matthew F. Brown, Pfizer Global Research and Development, Department of Antibacterials, Immunology, and Cancer, MS 8220-2410, Eastern Point Road, Groton, CT 06340, USA

Hal E. Broxmeyer, Dept Microbiology & Immunology, and Walther Oncology Center, Indiana University School of Medicine, 950 West Walnut Street, Indianapolis, IN 46202, USA; e-mail: hbroxmey@iupui.edu

Tassie L. Collins, Department of Neuroscience, Amgen Inc., 1120 Veterans Boulevard, South San Francisco, CA 94080, USA; e-mail: tassiec@amgen.com

Stephen W. Chensue, Department of Pathology, The University of Michigan Medical School, 1301 Catherine Road, Ann Arbor, MI 48109-0602, and The University of Michigan Medical and Veterans Administration Hospital, Ann Arbor, MI, USA

Nicole Frilot, Immunotherapy Center and Department of Pathology, Medical College of Georgia, Augusta, GA, USA

Nobutaka Fujii, Kyoto University, Kyoto, Japan

Ronald P. Gladue, Pfizer Global Research and Development, Department of Antibacterials, Immunology, and Cancer, MS 8220-2410, Eastern Point Road, Groton, CT 06340, USA; e-mail: Ronald.P.Gladue@Groton.Pfizer.com

Nuria Godessart, Almirall Research Center, Sant Feliu de Llobregat, Spain; e-mail: nuria.godessart@almirall.es

Paul J. Higgins, UCB Research/UCB Pharma, 840 Memorial Drive, Cambridge, MA 02139, USA; e-mail: paul.higgins3@comcast.net

Cory Hogaboam, Department of Pathology, The University of Michigan Medical School, 109 Zina Pitcher Place, Ann Arbor, MI 48109-0602, USA

Michael G. Johnson, Department of Chemistry, Amgen Inc., 1120 Veterans Boulevard, South San Francisco, CA 94080, USA

William J. Karpus, Department of Pathology, Northwestern University Feinberg School of Medicine, 303 E Chicago Avenue, W127, Chicago, IL 60611, USA; e-mail: w-karpus@northwestern.edu

Steven L. Kunkel, Department of Pathology, The University of Michigan Medical School, 4071BSRB, 109 Zina Pitcher Place, Ann Arbor, MI 48109-0602, USA; e-mail: slkunkel@umich.edu

Nicholas Lukacs, Department of Pathology, The University of Michigan Medical School, 109 Zina Pitcher Place, Ann Arbor, MI 48109-0602, USA

Andrew D. Luster, Center for Immunology and Inflammatory Diseases, Division of Rheumatology, Allergy and Immunology, Massachusetts General Hospital, Harvard Medical School, 149 Thirteenth Street, Room 8301, Charlestown, MA 02129, USA; e-mail: aluster@partners.org

Zamaneh Mikhak, Center for Immunology and Inflammatory Diseases, Division of Rheumatology, Allergy and Immunology, Massachusetts General Hospital, Harvard Medical School, 149 Thirteenth Street, Room 8301, Charlestown, MA 02129, USA

Julio C. Medina, Department of Chemistry, Amgen Inc., 1120 Veterans Boulevard, South San Francisco, CA 94080, USA

Jean-Marie Nicolas, UCB Research/UCB Pharma, 840 Memorial Drive, Cambridge, MA 02139, USA

Stephen C. Peiper, Immunotherapy Center and Department of Pathology, Medical College of Georgia, Augusta, GA, USA; e-mail: speiper@mcg.edu

Louis M. Pelus, Dept Microbiology & Immunology, and Walther Oncology Center, Indiana University School of Medicine, 950 West Walnut Street, Indianapolis, IN 46202, USA; e-mail: lpelus@iupui.edu

Christine A. Power, Molecular Biology, Serono Pharmaceutical Research Institute, 14, Chemin des Aulx, 1228 Plan-les Ouates, Switzerland;
e-mail: christine.power@serono.com

Amanda E. I. Proudfoot, Protein Biochemistry, Serono Pharmaceutical Research Institute, 14, Chemin des Aulx, 1228 Plan-les Ouates, Switzerland;
e-mail: amanda.proudfoot@serono.com

Shon R. Pulley, Eli Lilly and Company, Lilly Corporate Center, Discovery Chemistry & Research Technologies, Indianapolis, IN 46285, USA;
e-mail: pulley_shon_r@lilly.com

Henry M. Sarau, Respiratory and Inflammation CEDD, GlaxoSmithKline, P.O. Box 1539, King of Prussia, PA 19406, USA; e-mail: Henry.Sarau@gsk.com

C. Eric Schwartz, UCB Research/UCB Pharma, 840 Memorial Drive, Cambridge, MA 02139, USA

Matthias Schwarz, Serono Pharmaceutical Research Institute, 14, Chemin des Aulx, 1228 Plan-les Ouates, Switzerland

Hirokazu Tamamura, Kyoto University, Kyoto, Japan

Timothy N.C. Wells, Protein Biochemistry, Serono Pharmaceutical Research Institute, 14, Chemin des Aulx, 1228 Plan-les Ouates, Switzerland

Zi-xuan Wang, Immunotherapy Center and Department of Pathology, Medical College of Georgia, Augusta, GA, USA

John R. White, Biopharmaceutical CEDD, GlaxoSmithKline, P.O. Box 1539, King of Prussia, PA 19406, USA; e-mail: John.R.White@gsk.com

Preface

The discovery of interleukin-8 close to 20 years ago initiated a new field of research touching on many aspects of immunology and inflammation. Interleukin-8 is just one member of a large class of structurally related chemoattractant proteins, known as chemokines. Chemokines are involved in the traffic control of leukocytes, which bear the corresponding chemokine receptors on their surfaces. They are the largest family of cytokines in the human genome. The discovery of chemokines and chemokine receptor has been largely fueled by the human genome sequencing efforts. To date, there are more then 45 known chemokines and approximately 17 receptors.

Chemokine research over the last two decades has focused on their role in leukocyte migration. It is now clear that chemokines affect all aspects of immunology and contribute to the pathology of a large number of inflammatory and immune mediated diseases, such as rheumatoid arthritis, pulmonary inflammatory diseases and multiple sclerosis. Their fundamental contributions to chronic inflammatory diseases make them a principal target for the development of novel, anti-inflammatory therapeutics.

More recently, it has become apparent that chemokines have an essential role in diverse processes distinct from their function in immunity, including tumor cell growth and metastasis, atherosclerosis and angiogenesis. This book gives a state-of-the-art account of recent developments in this field in the form of summaries written by highly regarded experts.

Volume I is focused on basic principles and progress in chemokine biology. The emphasis is on the role of chemokines in leukocytes function and on their role in dendritic cell biology. In addition, chemokine receptor signaling and natural antagonism of the receptors is covered. Finally, aspects of chemokine biology as it pertain to endothelial cells and angiogenesis is discussed. It is also published in the book series Progress in Inflammation Research and is entitled *Chemokine Biology – Basic Research and Clinical Application. Volume I: Immunobiology of Chemokines* (2006, Birkhäuser, ISBN 978-3-7643-6825-8).

Volume II deals with issues related to the pathophysiology of chemokines, chemokine-related drug development and potential therapeutic applications. Chemokine targeted therapeutics are at a critical stage of clinical development. The clinical utility of chemokine receptor modulators cannot be predicted with assurance given the redundancies and overlapping biological activities of chemokines and the fact that multiple chemokines interactive with specific and overlapping chemokine receptor. Volume II presents the current status of two important chemokine receptor antagonists that are in the final phases of development: CCR5 antagonist and CXCR4 antagonist, and also provides a summary of other chemokine receptor modulators that are the next waves of potential therapeutics.

Both books provide both introductory and novel information for a broad readership, including clinicians and biomedical scientists.

December 2006

Kuldeep Neote
Gordon L. Letts
Bernhard Moser

Chemokines in animal disease models

Chemokines in animal models of inflammation

Steven L. Kunkel[1], *Nuria Godessart*[3], *Cory Hogaboam*[1], *Stephen W. Chensue*[1,2] *and Nicholas Lukacs*[1]

[1]Department of Pathology, The University of Michigan Medical School, 109 Zina Pitcher Place, Ann Arbor, MI 48109-0602, USA; [2]The University of Michigan Medical and Veterans Administration Hospital, Ann Arbor, MI, USA; [3]Almirall Research Center, Sant Feliu de Llobregat, Spain

Introduction

There is little doubt that animal modeling of various human diseases, disorders, and toxicology has been a staple of both academia and the pharmaceutical industry for decades. While it may be difficult to argue that experimental models, in particular rodent models, of human disease provide an exact mimic of their human counterpart, these experimental systems do provide a mechanism to generate meaningful data in the absence of human specimens. This is especially true in the case of human inflammatory disease where one is unlikely to gain access to tissue samples during the early initiation of a disease, and in the case of chronic disease where continuous, longitudinal samples are nearly impossible to obtain to assess the maintenance stage of chronic disease. Thus, the strategic use of experimental animal systems will continue to be an important tool in assessing both mechanisms of disease and efficacy of drugs which target these diseases.

The tactical use of animal models has been extremely important in understanding various aspects of inflammation; however, no area of inflammatory disease has benefited more from the use of experimental models than cytokine and chemokine biology. The collective use of knockout, knock-in, and transgenic animals, along with viral delivered genes, neutralizing antibodies, antisense oligonucleotides, aptamers, and siRNA have all targeted various inflammatory mediators in murine models in an attempt to further our understanding of disease mechanisms.

In the past decade there has been an explosion of data outlining the importance of individual chemokines to the initiation, maintenance, and resolution of inflammatory disease based on the use of well-developed animal models. These models have identified the importance of chemokine to both the innate immune response, as well as to the evolution of sophisticated acquired immunologic responses. The scientific community will continue to rely upon the use of experimental animal modeling to uncover the enigmatic mechanisms of chemokine biology.

Chemokine Biology – Basic Research and Clinical Application, Volume II
edited by Kuldeep Neote, Gordon L. Letts and Bernhard Moser

Experimental acute systemic inflammation (sepsis)

The clinical manifestations of sepsis are usually the consequence of an intense host immune response to either a known or uncharacterized insult [1–3]. Despite significant advances in intensive care unit technology, antibiotic development, and mechanical ventilatory support, mortality due to sepsis-mediated multiorgan failure has not changed significantly for decades [4]. This underscores the fact that pharmacologic options available to effectively manage these patients are often limited and reflect our limited knowledge of host-dependent mechanisms that drive these pathophysiologic disorders. An additional complicating factor is that timing is key; often by the time the syndrome is diagnosed the host has already responded with exaggerated mediator expression, alterations in physiology, tissue injury, and organ dysfunction. These features contribute to the rapid deterioration of the patient's health and have led physicians to define sepsis as a deadly acute disease. This loose definition of sepsis derives from clinical investigations that have studied short-term outcomes and report mortalities of 30–50%, irrespective of whether an infectious agent was identified [5, 6].

Dogma of the inflammatory response argues that the initial host response during sepsis relies upon an innate system for a programmed common pathway enacted against a variety of recognized pathogen patterns or signals, which may induce etiologic agents of sepsis. If this early innate reaction fails to localize and clear the challenge, systemic activation of the immune system likely occurs with high levels of pro- (systemic inflammation response syndrome – SIRS) followed by anti-inflammatory cytokines, system in compensatory anti-inflammatory response syndrome (CARS) resulting in host-derived cytokine storms. The consequence of unregulated cytokine production mechanistically contributes to the multiple clinical pathologies associated with sepsis, including immunosuppression (Fig. 1).

Using an experimental model of cecal ligation and puncture (CLP), mechanisms have been identified by which specific chemokines may act as key mediators linking the innate immune response to ensuing events after CLP [7]. Recent data would predict that specific chemokines may alter the morbidity/mortality of experimental sepsis by regulating the expression of toll-like receptors, cytokine phenotype, and lung macrophage and dendritic cell activity.

Chemokine expression in clinical and experimental sepsis

Clinical investigations have identified elevated levels of both CXC and CC chemokines associated with human sepsis and subsequent acute lung injury [8–10]. Data from clinical investigations demonstrate that levels of CCL2 significantly correlate with the disease course of sepsis [10]. The increased elevation in CCL2 did not correlate with leukocyte numbers, suggesting that the mononuclear cell chemo-

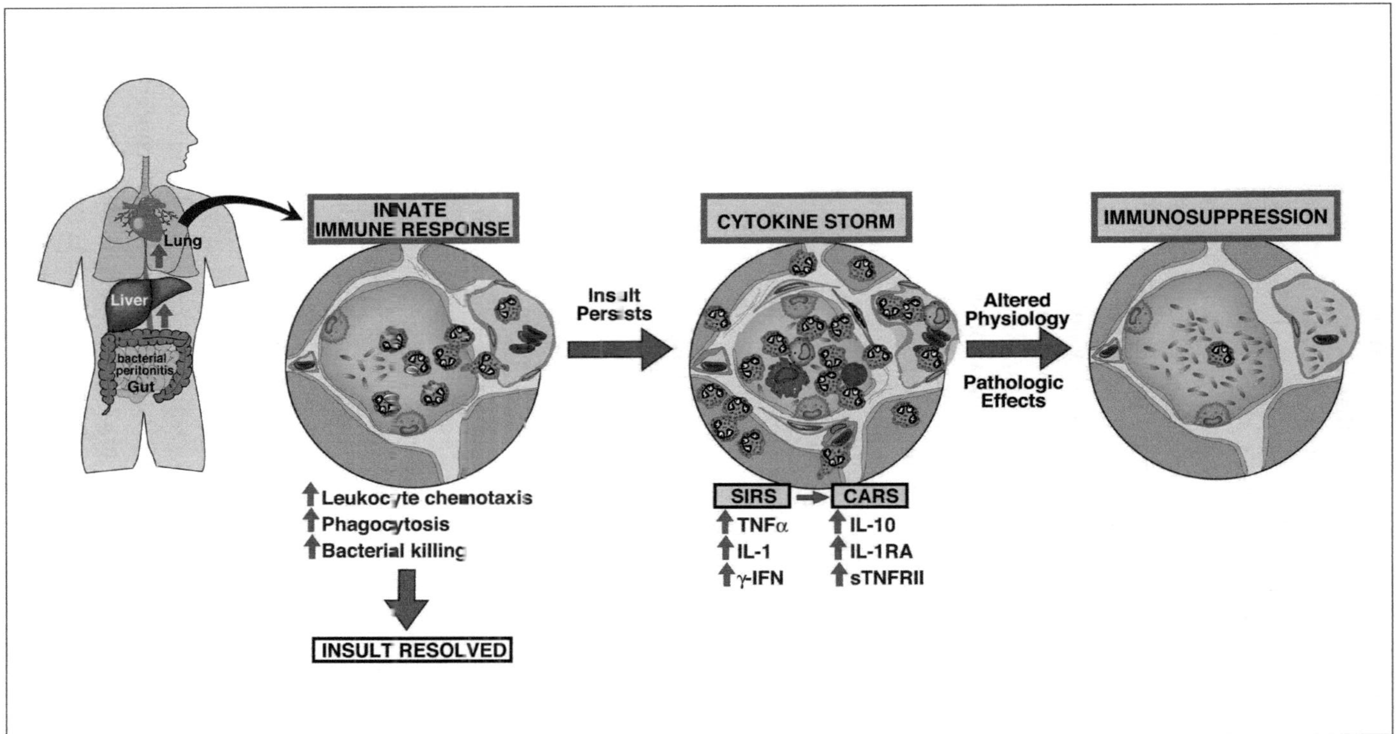

Figure 1
Both clinical and experimental sepsis is a disease process driven by a variety of pro-and anti-inflammatory mediators. Chemokines clearly play a role in this enigmatic process by serving as activators of the inflammatory process.

tactic activity may not have been the important biological action in this disease setting. This latter observation suggests that specific chemokines may have different biological activities during the evolution of sepsis.

This concept is supported by experimental data suggesting a compensatory response of chemokines, as the body attempts to reduce the deleterious effects of sepsis [11–13]. In these studies, animals receiving exogenous CCL2 were spared when challenged with a lethal dose of either lipopolysaccharide (LPS) or bacteria. One potential mechanism for the protection against the LPS challenge appears to be via a reduction in the production of IL-12 and TNF [12]. These data support the idea that specific chemokines may facilitate a shift from an inflammatory cytokine phenotype to a phenotype with compensatory activity. Additional information supports this concept by showing that other CC chemokines may be involved in regulating the maintenance of systemic inflammation associated with experimental sepsis. Studies have shown that the development of lethal endotoxemia is blocked in $CCR4^{-/-}$ mice and these knockout mice are resistant to a pulmonary challenge with *Aspergillus conidia* [14, 15]. Unpublished data demonstrates that passive immunization of mice with antibody directed against the CCR4 ligand CCL17 blocks the lethality of CLP. This suggests an important role for chemokines in the evolution and maintenance of experimental sepsis. An additional observation that ties CC chemokines to the innate immune response is the ability of CCL22 to augment macrophage function [16]. Studies have shown that CCL22 may activate the innate immune response, as treated macrophages have an increase in phagocytosis, bacterial killing, and an oxidative burst [16]. Collectively, these investigations support a paradigm where specific chemokines are intimately involved in the innate immune response of the host.

Chemokines provide a link between innate and adaptive immunity

The key to the successful operation of the host defense system is the culmination of a number of interactive processes, which place constant pressure on eliminating a foreign antigen or pathogen. Different levels of immune sophistication dictate these interactive processes of the immune/inflammatory response, ranging from the nonspecific reaction to pattern recognition molecules (PAMs) to the highly sophisticated specific cell-mediated acquired response to a single antigen. Innate immunity constitutes the former reaction and is triggered within minutes of exposure to a foreign agent. The latter response is characteristic of an acquired immune response and requires a number of days to imprint its effect on the host. A concerted and interactive innate and acquired immune reaction is key for an automatic, dynamic, sustained, and regulated response toward clearing a foreign agent. Thus, it is imperative that the *in vivo* concept of innate and acquired immunity be considered a synergistic longitudinal assault on a foreign agent not as separate entities.

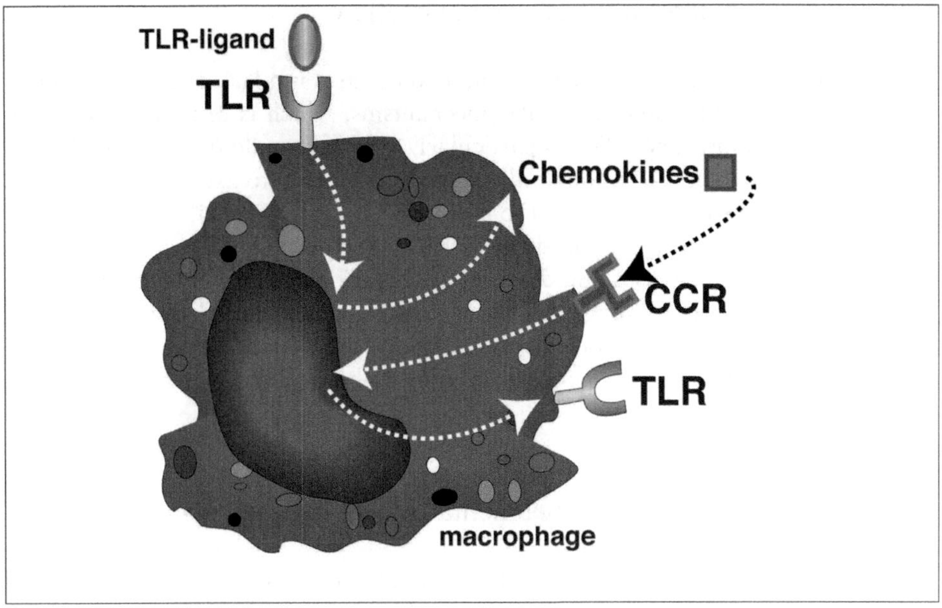

Figure 2
One of the consequences of activating leukocytes via toll-like receptors (TLR) is the generation of various chemokines. In turn, these chemokines are likely involved in a feedback process to regulate the expression of TLR.

The chemokine receptors CCR2, 4, 5 and 8 and their ligands have all clearly been associated with sophisticated aspects of acquired immunity, and in many instances with a highly refined type 2 response. However, these same chemokine receptors also are involved in early inflammatory responses, as they have been coupled to macrophage and neutrophil mediated innate activation events. For example, CCR2/CCL2 has been identified as key participants in type 2 polarization, as well as playing an important role in innate immunity as macrophage activators. The receptor for CCL22 and CCL17, CCR4, has been implicated in the recruitment of Th-2 cells. However, CCR4$^{-/-}$ mice are extremely resistant to acute endotoxin challenge (sepsis) and have a greatly increased ability to eliminate fungal challenge to the lung, as compared to wild type mice [14, 15]. This information underscores the functional activity of specific CC chemokines as they operate in very different spheres of the immune response. Thus, chemokines normally associated with chronic inflammation can also exert novel regulatory effects, as they actively participate in innate immune responses. This latter idea may be particularly true as chemokine expression and regulation may be intimately linked to the expression and activation of toll-like receptors (TLR) (Fig. 2).

Experimental granulomatous lung inflammation

A common theme of almost all chronic human diseases is a lack of a precise understanding of the cellular and molecular mechanisms, which is responsible for their evolution and maintenance. This is particularly true of granulomatous lung disease, which remains an enigma, as longitudinal studies designed to assess mechanisms of human lung granulomas are difficult to perform. Therefore, animal models designed to mimic specific aspects of the developing lung granuloma are crucial to both understanding these diseases and developing efficacious therapy. A number of animal models have been developed with a working hypothesis that the initiation, maintenance and final resolution of the reactions are dependent upon chemokine mediator systems that may be similar in both the human and experimental model systems.

Cytokines, chemokines and experimental interstitial lung inflammation

There is a growing body of scientific evidence suggesting that the profile of cytokines, chemokines, and their receptors, which are expressed during the evolution of an inflammatory response in the lung, are determining factors that lead to the chronicity of lung disease. Much of the supporting evidence that defines the mediators associated with chronic lung disorders is derived from historical studies demonstrating that IL-12, TNF and IFN-γ have profound effects on the evolution of cell-mediated inflammation induced by intracellular infectious agents, while IL-4, IL-5 and IL-13 appear to be dominant cytokines expressed during asthma or immune responses caused by certain extracellular infectious agents [17–22]. Furthermore, recent studies suggest that specific chemokines may be associated with the development of chronic immune responses in the lung with either type 1 or type 2 cytokine phenotypes. For example, CCL3 and activation of its receptor CCR1 have been implicated in the evolution of a type 1 process [23–25], while CCL2 and CCL17 and their respective receptors CCR2 and CCR4 have been associated with a type 2 response [26]. Thus, certain cytokine and chemokine profiles are in keeping with a general theme that certain chronic, cell-mediated responses possess a polarized cytokine expression bias.

CCL17, CCL22 and CCR4 in chronic lung inflammation

A number of published clinical and experimental studies provide support to the idea that CCR4 and its ligands, CCL17 and CCL22, serve as molecular mechanism for the pathologic consequences associated with the chronic lung inflammation [27–30]. While many of these investigations have assessed the expression of CCR4,

CCL17 and CCL22 by polarized T cells *in vitro*, others studies have identified their expression patterns associated with atopic asthmatics in human samples [31]. In these latter studies, immunofluorescent analysis of endobronchial biopsies from asthmatics challenged with allergen demonstrated that many of the tissue T cells had a type 2 cytokine phenotype and expressed CCR4, while the non-inflammatory structural cells in the lung produced CCL22 and CCL17 [31]. These studies are in keeping with the majority of presently published reports on CCR4, CCL22 and CCL17 that highlights their association with the evolution of an allergic response with a polarized cytokine phenotype; however, little is known regarding the role of CCR4, CCL17 and CCL22 during the evolution of other clinical and experimental models of chronic inflammation.

Preliminary data begins to address this deficiency and provide insight into the role that this receptor and its ligands may play in non-allergic models of chronic lung inflammation. Especially intriguing are studies assessing CCL17 immunoneutralization in animals with developing lung granulomatous inflammation. Data demonstrates that the removal of CCL17 reduces the cell-mediated immune response in the lung, changes the cellular composition of the lesion, and reduces collagen deposition at the site of granuloma development. These data corroborate very recent published findings where $CCL17^{-/-}$ mice failed to develop a normal skin delayed-type hypersensitivity (DTH) response and enhanced the survival of allografts, as compared to wild-type controls [32]. In additional studies, CCL17 expressing cells were found in abundance in the draining lung lymph nodes of animals with developing lung granulomas, but were never found in the spleen of these animals.

A similar report based on tracking CCL17 expressing cells with a green fluorescent protein marker found that CCL17 producing cells were found at regional lymph nodes but not in the spleen, even after systemic antigen challenge [32]. This data would support the concept that CCL17-dependent attraction of specific subpopulations of leukocytes is favored at regional sites of antigenic stimulation, but not systemic sites. Collectively, this information underscores the novel role of CCL17 during the evolution of inflammation and provides a strong rational to investigate the understudied mechanisms whereby CCR4 and its ligands contribute to the maintenance of chronic inflammatory lung disease.

Experimental asthma

Chemokines have diverse functions during allergic asthmatic responses, which relate to a number of interacting functions including recruitment, cellular activation/degranulation, differentiation, as well as directly altering the immune response. The identification of chemokines in the airways of asthmatics after allergen provocation likely demonstrates their importance during disease. Furthermore, the expression of distinct chemokine receptors on infiltrating cell populations, especially lymphocytes and eosinophils, provide excellent targets for therapeutic intervention. However, the identification of the exact receptor expression must be verified during the development of ongoing disease. To this end, animal models of asthma have provided a number of interesting chemokine receptor targets to examine. The models vary substantially with several labs using ovalbumin-induced airway responses while others utilize natural allergens to induce the airways disease. In general, the results from these models correspond to one another.

One of the first targets to be recognized, CCR3, is highly expressed on recruited eosinophils, the cell that is most often correlated with the severity of disease. Subsequently, CCR3 can be found on mast cells, basophils and subsets of Th2 type cells, giving more justification for targeting CCR3 to modulate the recruitment and/or activation of a number of cell populations involved in pathogenesis of asthma. Although positive data have been generated in allergic animals treated with anti-CCL11/eotaxin, examination of CCL11$^{-/-}$ mice did not demonstrated a clear role for this chemokine in the allergic airway responses [33]. However, CCR3$^{-/-}$ mice have reduced eosinophil recruitment and airway hyperreactivity, an effect that was repeated using anti-CCR3 antibody in the ovalbumin model of allergic disease [34]. Studies using specific CCR3 antagonists should give a much clearer indication of whether CCR3 is a target for intervention.

A number of chemokine receptors have been shown to be preferentially expressed on Th2 type lymphocytes, including CCR4 and CCR8 [35]. The association of these receptors with IL-4 producing lymphocytes has been identified in both *in vitro* using Th2 skewing conditions and *in vivo* in lymphocytes isolated from asthmatic patients. Thus, these two receptors make logical targets to inhibit recruitment of these potentially disease associated cells out of the airway during allergic disease. However, animal models of disease using specific receptor knockout animals have left open the interpretation of whether these will provide good targets for inhibiting asthmatic disease. In CCR4 deficient animals the attenuated response appears to be centered on an altered innate immune response that has been examined within endotoxin and anti-fungal responses, with little effect in allergen-induced models. The use of CCR8$^{-/-}$ mice has provided data indicating that airway physiology changes were not attenuated during allergen challenges in CCR8-deficient mice derived from three different laboratories [36]. However, one of these studies that used three different Th2 cytokine mediated models demonstrated an alteration in the eosinophil-associated inflammatory response [24]. These latter findings are supported by studies demonstrating the expression and function of CCR8 on eosinophils [37] and in studies blocking the CCR8 specific ligand, CCL1. More recent findings have suggested that CCR8 preferentially identifies an IL-10 producing CD4^{+} T cell subset.

The upregulation of other chemokine receptors has also been identified in animal models of asthmatic disease. The expression of CCR6 in the lungs of allergic animals suggested that this receptor may play a role in the development of disease. CCR6 has been primarily associated with immature dendritic cells (DC) that once activated lose CCR6 expression followed by CCR7 upregulation to localize the DC to the lymph nodes. Original studies with $CCR6^{-/-}$ mice using lymphocyte transfer techniques demonstrated an altered migration of $CD4^{+}$ lymphocytes to the lung suggesting that tissue specific migration was altered leading attenuation of disease development. Since CCR6 is found on multiple cell populations, including B cells, immature dendritic cells, and eosinophils, it may be important to include these interactions in any model of altered inflammation [38]. More recent studies have confirmed that altered T cell recruitment into the lung is one of the possible defects in the $CCR6^{-/-}$ mice [39]. However, these latter studies also found that the $CCR6^{-/-}$ mice also displayed defects in dendritic cell accumulation in the lungs of allergic mice demonstrating that multiple cell population defects within these mice.

Rheumatoid arthritis

The involvement of chemokines in the pathophysiology of rheumatoid arthritis (RA) is experimentally supported by studies demonstrating the presence of high levels of chemokines in the synovial fluid, and the expression of chemokine receptors in cells infiltrating the synovium of arthritic joints [40, 41]. These studies showed that multiple chemokines of the CC (CCL2, CCL3, CCL5), CXC (CXCL1-5, CXCL8, CXCL12) and CX3C (CX3CL1) families, are potentially involved in the regulation of, at least, two relevant processes: cell recruitment and angiogenesis [42]. Furthermore, studies analyzing the impact of several antiarthritic therapies such as infliximab, leflunomide, methotrexate or sulfasalazine on chemokine expression have reported that their efficacy is associated with a reduction in both chemokine expression and cell infiltration [43–47]. Some authors have postulated that circulating levels of CCL2 and CCL5 could be used as markers of joint inflammation and radiological progression, respectively, in RA [46, 47].

Animal models of arthritis do not reproduce all the features of human rheumatoid arthritis, but they help to delineate the inflammatory and immune processes that take place in the course of the disease. Several rodent models of RA are available; ranging from the spontaneous disease developed in MLR/Fas^{lpr} mice, to the induction of arthritis following the injection of complete Freund's adjuvant (CFA), collagen II, streptococcal cell wall or mBSA to mice and rats [48]. Depending on the rodent species used, the inducing agent, and the administration protocol (that is, if treatment starts before or after the onset of the disease), results obtained may differ substantially. The most extensively used models of arthritis are the collagen II-induced arthritis (CIA) in DBA-1 mice, and the adjuvant-induced arthritis in Lewis or Wistar rats. The characterization of the temporal expression of chemokines in these two models has been reported [42, 49]. In the last decade, the explosion of genetically-modified mice, together with the higher availability of tools (antibodies, ELISA kits) for mouse studies, have favored the use of the murine model.

Several studies performed in the last 5 years have demonstrated, using truncated chemokines (functional receptor antagonists) or anti-chemokine antibodies, that the blockade of chemokines like CCL2, CCL3, CCL5, CX3CL1 or CXCL13, improves the course of experimental arthritis [42, 50, 51]. Using small molecule compounds or truncated chemokines, it has been suggested that single antagonism of CCR1 or CXCR4, and dual antagonism of CCR5/CXCR3 or CCR1/CCR5 could be effective to treat arthritis [52–55]. Intriguingly, the first results of experimental arthritis in chemokine or chemokine receptor knockout mice were not reported until 2004 [56]. The experimental complication that represents backcrossing the knockout mice with the DBA-1 mice for 6–8 generations in order to get animals susceptible to CIA could in part explain this fact.

Studies of CIA induction in CCR2- and CCR5-deficient mice on a DBA-1 background have shown that, whereas the lack of CCR5 had no impact in the progression of the disease, CCR2-null mice developed a much worse arthritis than controls. These results suggested that CCR2 had a protective role in arthritis. As it has been described in other disease models, the lack of CCR2 in arthritis was found to be dispensable for the migration of monocytes to the inflamed joints [56]. Another group has studied the impact of CCR2 blockade in early and late phases of the disease using an anti-CCR2 antibody [57]. The administration of the antibody during the initiation of the disease (days 1–15) improved clinical signs of arthritis and histological scores measured as leukocyte infiltration, synovial hyperplasia, and bone and cartilage erosion. However, when the antibody was administered from days 21–36, it markedly aggravated clinical and histological signs of arthritis and increased the humoral immune response against collagen. The presence of CCR2 in regulatory T cells could explain the role of CCR2 in down-modulating the inflammatory response [57].

Unfortunately, results indicating whether studies of chemokine receptor neutralization in animal models translate well into the human disease are lacking. Compound CP-481,715, a CCR1 antagonist from Pfizer, recently failed to demonstrate efficacy in a 6-week clinical trial in patients with active RA [58]. No results of the effect of this antagonist in animal models of RA have been published, probably because of the lack of affinity of the compound for the rodent receptors. Moreover, two agents blocking CCR2 (a small molecule compound and an antibody) have progressed to Phase II clinical trials for RA. Since CCR2 has been devalidated in CIA, the results emerging from the human trials will teach us about *in vivo* target validation in the chemokine field.

Conclusion

While a significant amount of finances and scientific energy has been invested in targeting chemokine biology in experimental disease, no clinical validation of chemokine receptors as a therapeutic target exist yet. Provocative investigations have been reported for the involvement of chemokines in diverse experimental disease models, which continue to serve as enticing segues for drug development. The constant search for small-molecule receptor antagonists is ongoing, and other strategies targeting chemokines will undoubtedly be developed in the future. In the next few years, we will know if these drugs have a place in the treatment of autoimmune and chronic, inflammatory disease, AIDS or cancer.

Acknowledgements
These studies were supported in part by NIH grants #31237, 35276, 74024, 31963.

References

1 Rinaldo JER, Rogers RM (1982) Adult respiratory distress syndrome: Changing concepts of lung injury and repair. *N Engl J Med* 306: 900–909
2 Strieter RM, Lynch JP, Basha MA, Standiford TJ, Kasahara K, Kunkel SL (1990) Host responses in mediating sepsis and adult respiratory distress syndrome. *Seminars in Respir Infect* 5: 233–247
3 Angele MK, Faist E (2002) Clinical review: Immunodepression in the surgical patient and increased susceptibility to infection. *Crit Care* 6(4): 298–305
4 Mortelliti MP, Manning HL (2002) Acute respiratory distress syndrome. *Am Fam Physician* 65(9): 1823–1830
5 Perl TM, Dvorak L, Hwang T, Wenzel RP (1995) Long-term survival and function after suspected gram-negative sepsis. *JAMA* 274: 338–345
6 Quartin AA, Schein RM, Kett DH, Peduzzi PN (1997) Magnitude and duration of the effect of sepsis on survival. *JAMA* 277: 1058–1063
7 Steinhauser ML, Hogaboam CM, Lukacs NW, Strieter RM, Kunkel SL (1999) Multiple roles for IL-12 in a model of acute septic peritonitis. *J Immunol* 162: 5437–5433
8 Donnelly SC, Strieter RM, Kunkel SL, Walz A, Steedman D, Grant IS, Pollok AJ, Carter DC, Haslett C (1994) Chemotactic cytokines in the established adult respiratory distress syndrome and at-risk patients. *Chest* 105(3 Suppl): 98S–99S
9 Donnelly SC, Strieter RM, Kunkel SL, Walz A, Robertson CR, Carter DC, Grant IS, Pollok AJ, Haslett C (1993) Interleukin-8 and development of adult respiratory distress syndrome in at-risk patient groups. *Lancet* Mar 341(8846): 643–647
10 Bossink AW, Paemen L, Jansen PM, Hack CE, Thijs LG, Van Damme J (1995) Plasma

levels of the chemokines monocyte chemotactic proteins-1 and -2 are elevated in human sepsis. *Blood* 86(10): 3841–3847

11 Matsukawa A, Hogaboam CM, Lukacs NW, Lincoln PM, Strieter RM, Kunkel SL (1999) Endogenous monocyte chemoattractant protein-1 (MCP-1) protects mice in a model of acute septic: cross-talk between MCP-1 and leukotriene B4. *J Immunol* 163(11): 6148–6154

12 Matsukawa A, Hogaboam CM, Lukacs NW, Lincoln PM, Strieter RM, Kunkel SL (2000) Endogenous MCP-1 influences systemic cytokine balance in a murine model of acute septic peritonitis. *Exp Mol Pathol* 68(2): 77–84

13 Matsukawa A, Hogaboam C, Lukacs NW, Lincoln P, Evanoff H, Kunkel SL (2000) Pivotal role of the CC chemokine, macrophage derived chemokine, in the innate immune response. *J Immunol* 164: 5362–5368

14 Schuh JM, Power CA, Proudfoot AE, Kunkel SL, Lukacs NW, Hogaboam CM (2002) Airway hyperresponsiveness, but not airway remodeling, is attenuated during chronic pulmonary allergic responses to Aspergillus in $CCR4^{-/-}$ mice. *FASEB J* 16(10): 1313–1315

15 Chvatchko Y, Hoogewerf AJ, Meyer A, Alouani S, Juillard P, Buser R, Conquet F, Proudfoot AE, Wells TN, Power CA (2000) A key role for CC chemokine receptor 4 in lipopolysaccharide-induced endotoxic shock. *J Exp Med* 191(10): 1755–1764

16 Matsukawa A, Hogaboam C, Lukacs NW, Lincoln P, Evanoff H, Kunkel SL (2000) Pivotal role of the CC chemokine, macrophage derived chemokine, in the innate immune response. *J Immunol* 164: 5362–5368

17 Romagnani S (1992) Induction of Th1 and Th2 responses: a key role for the natural immune response? *Immunol Today* 13: 379–381

18 Chensue SW, Warmington KS, Ruth J, Lincoln PM, Kunkel SL (1994) Cross-regulatory role of interferon-gamma (IFN), IL-4 and IL-10 in schistosome egg granuloma formation: *in vivo* regulation of Th activity and inflammation. *Clin Exp Immunol* 98: 395–400

19 Li YJ, Petrofsky M, Bermudez LE (2002) *Mycobacterium tuberculosis* uptake by recipient host macrophages is influenced by environmental conditions in the granuloma of the infectious individual and is associated with impaired production of interleukin-12 and tumor necrosis factor alpha. *Infect Immun* 70: 6223–6230

20 Fenhalls G, Stevens L, Bezuidenhout J, Amphlett GE, Duncan K, Bardin P, Lukey PT (2002) Distribution of IFN-gamma, IL-4 and TNF-alpha protein and CD8 T cells producing IL-12p40 mRNA in human lung tuberculous granulomas. *Immunology* 105: 325–335

21 Jakubzick C, Choi ES, Kunkel SL, Joshi BH, Puri RK, Hogaboam CM (2003) Impact of interleukin-13 responsiveness on the synthetic and proliferative properties of Th1- and Th2-type pulmonary granuloma fibroblasts. *Am J Pathol* 162: 1475–1486

22 Mountford AP, Coulson PS, Cheever AW, Sher A, Wilson RA, Wynn TA (1999) Interleukin-12 can directly induce T-helper 1 responses in interferon-gamma (IFN-gamma)

receptor-deficient mice, but requires IFN-gamma signalling to downregulate T-helper 2 responses. *Immunology* 97: 588–594

23 Lukacs NW, Kunkel SL, Strieter RM, Warmington K, Chensue SW (1993) The role of macrophage inflammatory protein 1 alpha in *Schistosoma mansoni* egg-induced granulomatous inflammation. *J Ex Med* 177: 1551–1559

24 Shang X, Qiu B, Frait KA, Hu JS, Sonstein J, Curtis JL, Lu B, Gerard C, Chensue SW (2000) Chemokine receptor 1 knockout abrogates natural killer cell recruitment and impairs type-1 cytokines in lymphoid tissue during pulmonary granuloma formation. *Am J Pathol* 157: 2055–2063

25 Gao JL, Wynn TA, Chang Y, Lee EJ, Broxmeyer HE, Cooper S, Tiffany HL, Westphal H, Kwon-Chung J, Murphy PM (1997) Impaired host defense, hematopoiesis, granulomatous inflammation and type 1-type 2 cytokine balance in mice lacking CC chemokine receptor 1. *J Exp Med* 185: 1959–1968

26 Warmington KS, Boring L, Ruth JH, Sonstein J, Hogaboam CM, Curtis JL, Kunkel SL, Charo IR, Chensue SW (1999) Effect of C-C chemokine receptor 2 (CCR2) knockout on type-2 (schistosomal antigen-elicited) pulmonary granuloma formation: analysis of cellular recruitment and cytokine responses. *Am J Pathol* 154: 1407–1416

27 Sekiya T, Yamada H, Yamaguchi M, Yamamoto K, Ishii A, Yoshie O, Sano Y, Morita A, Matsushima K, Hirai K (2002) Increased levels of a TH2-type CC chemokine thymus and activation-regulated chemokine (TARC) in serum and induced sputum of asthmatics. *Allergy* 57: 173–177

28 Nouri-Aria KT, Wilson D, Francis JN, Jopling LA, Jacobson MR, Hodge MR, Andrew DP, Till SJ, Varga EM, Williams TJ et al (2002) CCR4 in human allergen-induced late responses in the skin and lung. *Eur J Immunol* 32: 1933–1938

29 Hirata H, Arima M, Cheng G, Honda K, Fukushima F, Yoshida N, Eda F, Fukuda T (2003) Production of TARC and MDC by naive T cells in asthmatic patients. *J Clin Immunol* 23: 34–45

30 Leung TF, Wong CK, Lam CW, Li AM, Ip WK, Wong GW, Fok TF (2003) Plasma TARC concentration may be a useful marker for asthmatic exacerbation in children. *Eur Respir J* 21: 616–620

31 Panina-Bordignon P, Papi A, Mariani M, Di Lucia P, Casoni G, Bellettato C, Buonsanti C, Miotto D, Mapp C, Villa A et al. (2001) The C-C chemokine receptors CCR4 and CCR8 identify airway T cells of allergen-challenged atopic asthmatics. *J Clin Invest* 107: 1357–1364

32 Alferink J, Lieberam I, Reindl W, Behrens A, Weiss S, Huser N, Gerauer K, Ross R, Reske-Kunz AB, Ahmad-Nejaad P et al (2003) Compartmentalized production of CCL17 *in vivo*: strong inducibility in peripheral dendritic cells contrasts selective absence from the spleen. *J Exp Med* 197: 585–599

33 Rothenberg ME, MacLean JA, Pearlman E, Luster AD, and Leder P (1997) Targeted disruption of the chemokine eotaxin partially reduces antigen-induced tissue eosinophilia. *J Exp Med* 185: 785

34 Humbles AA, B Lu, Friend DS, Okinaga S, Lora, Al-Garawi, Martin TR, Gerard NP, Gerard C (2002) The murine CCR3 receptor regulates both the role of eosinophils and mast cells in allergen-induced airway inflammation and hyperresponsiveness. *Proc Natl Acad Sci USA* 99: 1479

35 Sallusto F, Mackay CR, Lanzavecchia A (2000) The role of chemokine receptors in primary, effector, and memory immune responses. *Annu Rev Immunol* 18: 593

36 Chung CD, Kuo F, Kumer J, Motani AS, Lawrence CE, Henderson WR Jr, Venkataraman C (2003) CCR8 is not essential for the development of inflammation in a mouse model of allergic airway disease. *J Immunol* 170: 581

37 Borchers MT, Ansay T, DeSalle R, Daugherty BL, Shen H, Metzger M, Lee NA, Lee JJ (2002) *In vitro* assessment of chemokine receptor-ligand interactions mediating mouse eosinophil migration. *J Leukoc Biol* 71: 1033

38 Lukacs NW, Prosser D, Wiekowski M, Lira SA, Cook DN (2001) Requirement for the chemokine receptor CCR6 in allergic pulmonary inflammation. *J Exp Med* 194: 551

39 Lundy SK, Lira SA, Smit JJ, Cook DN, Berlin AA, Lukacs NW (2005) Attenuation of allergen-induced responses in CCR6-/- mice is dependent upon altered pulmonary T lymphocyte activation. *J Immunol* 174: 2054

40 Godessart N, Kunkel SL (2001) Chemokines in autoimmune disease. *Curr Opin Immunol* 13: 670–675

41 Kunkel SL, Godessart N (2002) Chemokines in autoimmunity: from pathology to therapeutics. *Autoimmun Rev* 1: 313–320

42 Koch AE (2005) Chemokines and their receptors: future targets? *Arthritis Rheum* 52: 710–721

43 Nissinen R, Leirisalo-Repo M, Peltomaa R, Palosuo T, Vaarala O (2004) Cytokine and chemokine receptor profile of peripheral blood mononuclear cells during treatment with infliximab in patients with active rheumatoid arthritis. *Ann Rheum Dis* 63: 681–687

44 Ho CY, Wong CK, Li EK, Tam LS, Lam CW (2003) Suppressive effect of combination treatment of leflunomide and methotrexate on chemokine expression in patients with rheumatoid arthritis. *Clin Exp Immunol* Jul 133(1): 132–138

45 Volin MV, Campbell PL, Connors MA, Woodruff DC, Koch AE (2002) The effect of sulfasalazine on rheumatoid arthritic synovial tissue chemokine production. *Exp Mol Pathol* 73: 84–92

46 Ellimngsen T, Buus A, Syengarrd-Petersen K (2001) Plasma monocyte chemoattractant protein 1 is a marker for joint inflammation in rheumatoid arthritis. *J Rheumatol* 28: 41–46

47 Boiardi L, Macchioni P, Meliconi R, Pulsatelli L, Facchini A, Salvarani C (1999) Relationship between serum RANTES levels and radiological progression in rheumatoid arthritis patients treated with methotrexate. *Clin Exp Rheumatol* 17: 419–425

48 Bendele A (2001) Animal models of rheumatoid arthritis. *J Musculoskelet Neuronal Interact* 1: 377–385

49 Kasama T, Strieter RM, Lukacs NW, Lincoln PM, Burdick MD, Kunkel SL (1995) Inter-

leukin 10 expression and chemokine regulation during the evolution of murine type II collagen-induced arthritis. *J Clin Invest* 95: 2868–2876

50 Nanki T, Urasaki Y, Nishimura M, Muramot K, Kubota T, Miyasaka N (2004) Inhibition of fractalkine ameliorates murine collagen-induced arthritis. *J Immunol* 173: 7010–7016

51 Zheng B, Ozen Z, Zhang X, De Silva S, Marinova E, Guo L, Wansley D, Huston DP, West MR, Han S (2005) CXCL13 neutralization reduces the severity of collagen-induced arthritis. *Arthritis Rheum* 52: 620–626

52 Saeki T, Naya A (2003) CCR1 chemokine receptor antagonist. Current Pharm Design 9: 1201–1208

53 Matthys P, Hatse S, Vermeire K, Wuyts A, Bridger G, Henson GW, De Clercq E, Billiau A, Schols D (2001) AMD3100, a potent and specific antagonist of the stromal cell-derived factor-1 chemokine receptor CXCR4, inhibits autoimmune joint inflammation in IFN-gamma receptor-deficient mice. *J Immunol* 167(8): 4686–4692

54 Yang YF, Mukai T, Gao P, Yamaguchi N, Ono S, Iwaki H, Obika S, Imanishi T, Tsujimura T, Hamaoka T et al. (2002) A non-peptide CCR5 antagonist inhibits collagen-induced arthritis by modulating T cell migration without affecting anti-collagen T cell responses. *Eur J Immunol* 32: 2124–2132

55 Plater-Zyberk C, Hoogewerf AJ, Proudfoot AE, Power CA, Wells TN (1997) Effect of a CC chemokine receptor antagonist on collagen induced arthritis in DBA/1 mice. *Immunol Lett* 57(1–3): 117–120

56 Quinones MP, Ahuja SK, Jimenez F, Schaefer J, Garavito E, Rao A, Chenaux G, Reddick RL, Kuziel WA, Ahuja SS (2004) Experimental arthritis in CC Chemokine receptor 2-null mice closely mimics severe human rheumatoid arthritis. *J Clin Invest* 113: 856–866

57 Bruhl H, Cihak J, Schneider MA, Plachy J, Rupp T, Wenzel I, Shakarami M, Milz S, Ellwart JW, Stangassinger M et al (2004). Dual role of CCR2 during initiation and progression of collagen-induced arthritis: evidence for regulatory activity of $CCR2^+$ T cells. *J Immunol* 172: 890–898

58 Gladue RP, Zwillich SH, Clucas AT, Brown MF (2004) CCR1 antagonists for the treatment of autoimmune diseases. *Curr Opin Invest Drugs* 5: 499–504

Autoimmune diseases

William J. Karpus

Department of Pathology, Northwestern University Feinberg School of Medicine,
303 E Chicago Avenue, W127, Chicago, IL 60611, USA

Introduction

Experimental models of autoimmune disease have been used to dissect the mechanisms of disease pathogenesis in the corresponding human diseases. This chapter will deal with experimental autoimmune encephalomyelitis (EAE) as a model for human multiple sclerosis (MS) and experimental autoimmune diabetes (EAD) in the NOD mouse as a model for human diabetes. In the case of these tissue-specific autoimmune diseases, the autoreactive lymphocytes originate in lymph nodes and must migrate to either the central nervous system (CNS) in the case of EAE or the pancreas in the case of EAD. The accepted paradigm of leukocyte migration from blood into tissue involves a number of molecular events including selectin binding, chemokine binding, integrin binding and activation, and extravasation [1, 2]. Therefore, chemokines have a central role in the pathogenesis of tissue-specific autoimmune diseases. The approaches that have been used to study the role of chemokines in animal models of autoimmune disease included assessing tissue-specific temporal chemokine expression patterns, assessing corresponding chemokine receptor expression patterns on the tissue-infiltrating leukocytes, using neutralizing anti-chemokine therapy, employing chemokine and/or chemokine receptor knockout mice in the various disease models, and making transgenic mice that overexpress certain chemokines in specific tissues. This has resulted in the identification of subsets of the chemokine and chemokine receptor families that play a role in disease pathogenesis. This chapter will review the role of chemokines and their receptors in EAE and EAD as examples of tissue-specific autoimmune diseases.

Experimental autoimmune encephalomyelitis

EAE is a CD4$^+$ T cell-mediated, CNS demyelinating disease that serves as a model for the study of MS and its underlying pathophysiologic mechanisms [3]. Several reports have demonstrated an association between chemokine mRNA or protein

Chemokine Biology – Basic Research and Clinical Application, Volume II
edited by Kuldeep Neote, Gordon L. Letts and Bernhard Moser

expression and appearance of clinical disease [4–6]. Hulkower et al. [7] were the first to demonstrate the correlation between chemokine expression and development of EAE in the Lewis rat model. Subsequently, Ransohoff et al. [8] described expression of chemokine mRNA in the CNS of SJL/J mice with relapsing EAE. Using semi-quantitative RT-PCR and *in situ* hybridization they demonstrated that CXCL10 and CCL2 were expressed in the spinal cord. Additional studies of relapsing EAE demonstrated upregulation of mRNA chemokine expression for CCL5, CCL4, CCL3, CCL1, CXCL10, CCL2, CXCL1, and CCL7 just prior to the first appearance of clinical symptoms in a mouse model of EAE and that the chemokine levels remained elevated throughout the course of the disease [5]. In addition to these chemokines, CCL6 expression has also been associated with EAE [9]. The evidence that chemokines are tightly associated with the induction of disease is that CNS chemokine mRNA expression correlates with histological signs of inflammation as expression is not detected in the absence of leukocyte infiltration [4, 10]. Colocalization experiments have shown that CCL3 and CCL5 were expressed by infiltrating leukocytes, while CXCL10 and CCL2 were expressed only by astrocytes [11]. In addition to the association between CNS mRNA levels and tissue-specific inflammation, CNS chemokine protein levels have been associated with differential phases of relapsing disease. CCL3 and CXCL10 protein levels have been shown to be elevated in the CNS following adoptive transfer of activated neuroantigen specific T cells [12–14] and correlate with acute disease development while CCL2 levels increase with the development of the relapsing phase of disease [6]. More recently CCL20 has been shown to affect the development of EAE, not by modulating CNS lymphocyte accumulation, rather, by regulating the induction of the auto reactive T cells [15]. It should be emphasized that the emerging data suggests different chemokine expression patterns in different EAE models and also in different mouse strains [16]. The biological importance of CNS chemokine expression in EAE as been demonstrated by two approaches: *in vivo* anti-chemokine antibody treatments or chemokine knockout mice. Anti-CCL3 [12] and anti-CXCL10 [13, 14] treatment prevented acute clinical EAE while anti-CCL2 treatment was shown to prevent relapsing disease [6]. In addition to the role for CCL2 in relapsing EAE, through the use of knockout mice this chemokine has also been shown to be important for CNS monocyte accumulation during acute clinical disease [17]. A significant finding from the *in vivo* neutralization studies is that while a wide variety of chemokines may be expressed during inflammatory autoimmune disease, only a subset of chemokines actually plays a significant biologic role in disease pathogenesis. In addition to regulating the migration/accumulation of leukocytes in the CNS during disease development and progression, chemokines also appear to regulate the trafficking of antigen presenting cells necessary to prime the autoreactive T cell response [15]. In rare instances the results from *in vivo* chemokine neutralization studies [13] do not agree with similar studies performed using chemokine deficient mice [18] indicating a potential compensatory mechanism involving chemokine ligands that bind the same receptor. Despite overwhelming evidence that chemokines are involved in leukocyte migration to the CNS during EAE resulting in paralytic disease development, tissue-specific transgenic expression of CCL2 can result in disease amelioration [19] due to the ability of this chemokine to regulate the Th1/Th2 differentiation program [20, 21].

Similar to the relationship between chemokines and EAE development, chemokine receptor mRNA analysis has been utilized to demonstrate expression in the CNS [22–25] with the general conclusion that as inflammation ensues, there is an accumulation of inflammatory cells bearing chemokine receptors. Correspondingly, a reduction in CNS inflammation results in less chemokine receptor mRNA expression. A number of recent studies using genetically modified mice have shown that CCR1 [26] and CCR2 [27] expression are biologically important for the development of acute EAE. In the CCR1 knockout mice there was approximately a 50% decrease in clinical disease severity; however, the mechanism behind disease attenuation is not known. Since both T cells and monocytes have been shown to express CCR1 [28], it is possible that CCR1 expression by either lymphocytes or monocytes or perhaps both is required for EAE development. In the CCR2 knockout mice, there was almost a total absence of disease due to a failure of monocytes, and not T cells, to traffic to the CNS [27, 29]. These two examples are in contrast to EAE induction in CCR5 knockout mice where the same level of disease severity was seen compared to wild type control animals [30]. Furthermore, depending on the disease induction protocol, the necessity for certain chemokine receptor expression can be bypassed [31]. An advance that has come from both the chemokine and chemokine receptor studies in EAE is development of small molecular weight antagonists to chemokine receptors. Indeed, a small molecular weight antagonist of CCR1 has shown efficacy in the inhibition of clinical EAE [32, 33].

Experimental autoimmune diabetes

Much of the research delineating the pathogenic mechanisms of human autoimmune diabetes has been gleaned from experimental models of spontaneous and induced disease. In the model of spontaneous autoimmune diabetes the NOD mouse develops a peri-islet insulitis that is followed by destruction of the islets and subsequent diabetes [34]. In induced models of autoimmune diabetes a viral epitope is expressed as a transgene in the islets and the mice are subsequently infected with the virus to develop insulitis and diabetes or pancreas is infected with lymphocytic choriomeningitis virus and disease ensues [35]. In both types of EAD a hallmark feature is chemokine-dependent migration of T cells to the pancreatic islets, accumulation of these cells, and self-tissue destruction [36, 37]. The pancreatic beta cells in the islets have been shown to express CXCL10 and CCL20 in pre-diabetic mice [38] while Th1 cells infiltrating into the pancreatic islets have been shown to express

CCL2, CCL3, CCL4, CCL5, CCL7, CCL12 CCL17, and CXCL10 [39–43]. CCL21 has also been shown to be expressed in the pancreas and required for CD8 T cell homing [44]. Despite the observations that a subset of chemokines are expressed in the pancreas prior to disease development, chemokine expression alone does not confer overt diabetes as demonstrated using an islet-specific CCL2 transgenic mouse [45]. Functional studies where CXCL10 [46], CCL3 [40], or CCL22 [42] was neutralized *in vivo* demonstrated a decrease in disease. Moreover, mice deficient for CCL3 also showed a decrease in disease development further indicating the importance of this chemokine in pathogenesis [40]. A variety of chemokine receptors specific for the aforementioned ligands including CCR5 [40], CXCR3 [47], and CCR4 [42] have been shown to be expressed by islet-infiltrating T cells. Experiments using genetically modified mice lacking these individual receptors have demonstrated less disease compared to controls. Additionally, transgenic expression of CCL22 in the pancreas resulted in the chemoattraction of CCR4-bearing T cells and acceleration of clinical disease [42]. Therefore, it appears that EAD pathogenesis is also controlled in part by chemokine and chemokine receptor expression.

Summary

Chemokines and their receptors are a family of inflammatory molecules that are associated with many tissue-specific inflammatory events and the CNS and pancreas are no exception. One current view of chemokines is to regulate the migration and/or accumulation of leukocytes at a particular tissue site for the general function of infection clearance and tissue repair. However, aberrant accumulation of leukocytes, including antigen-specific T cells and monocytes, can induce pathology and result in tissue-specific autoimmune and/or inflammatory disease. Our greatest understanding of the role of chemokines in CNS disorders comes from the EAE model where the temporal and spatial chemokine expression patterns appear to regulate mononuclear cell accumulation and subsequent disease development [16]. In the case of autoimmune disease or bystander inflammatory disease, it would be beneficial to limit the biological effect of chemokine expression in order to limit the extent of self-tissue damage. To this end small molecular weight chemokine receptor antagonists have been developed and are being evaluated for efficacy in disease models as well as human disease [48]. Indeed, a CCR1 antagonist that has demonstrated efficacy in a rat model of EAE [49] is being tested for efficacy in clinical trials with MS patients, demonstrating that understanding the chemokine-regulated events in animal models of human disease can result in translation to new clinical approaches. Therefore, understanding the role of the chemokine superfamily of inflammatory molecules in diseases of the CNS and pancreas will shed light on specific pathogenetic mechanisms as well as provide new targets for therapeutic intervention.

Acknowledgements
The author's work is supported in part by National Institutes of Health grants AI35934, NS34510, and NS023349.

References

1 Butcher EC (1992) Leukocyte-endothelial cell adhesion as an active, multi-step process: a combinatorial mechanism for specificity and diversity in leukocyte targeting. *Adv Exp Med Biol* 323: 181–194

2 Springer TA (1994) Traffic signals for lymphocyte recirculation and leukocyte emigration: the multistep paradigm. *Cell* 76: 301–314

3 Segal BM (2003) Experimental Autoimmune encephalomyelitis: cytokines, effector T cells, and antigen-presenting cells in a prototypical Th1-mediated autoimmune disease. *Curr Allergy Asthma Rep* 3: 86–93

4 Glabinski AR, Tani M, Tuohy VK, Tuthill RJ, Ransohoff RM (1995) Central nervous system chemokine mRNA accumulation follows initial leukocyte entry at the onset of acute murine experimental autoimmune encephalomyelitis. *Brain, Behavior, & Immunity* 9: 315–330

5 Godiska R, Chantry D, Dietsch GN, Gray PW (1995) Chemokine expression in murine experimental allergic encephalomyelitis. *J Neuroimmunol* 58: 167–176

6 Kennedy KJ, Strieter RM, Kunkel SL, Lukacs NW, Karpus WJ (1998) Acute and relapsing experimental autoimmune encephalomyelitis are regulated by differential expression of the CC chemokines macrophage inflammatory protein-1α and monocyte chemotactic protein-1. *J Neuroimmunol* 92: 98–108

7 Hulkower K, Brosnan CF, Aquino DA, Cammer W, Kulshrestha S, Guida MP, Rapoport DA, Berman JW (1993) Expression of CSF-1, c-fms, and MCP-1 in the central nervous system of rats with experimental allergic encephalomyelitis. *J Immunol* 150: 2525–2533

8 Ransohoff RM, Hamilton TA, Tani M, Stoler MH, Shick HE, Major JA, Estes ML, Thomas DM, Tuohy VK (1993) Astrocyte expression of mRNA encoding cytokines IP-10 and JE/MCP-1 in experimental autoimmune encephalomyelitis. *FASEB J* 7: 592–600

9 Asensio VC, Lassmann S, Pagenstecher A, Steffensen SC, Henriksen SJ, Campbell IL (1999) C10 is a novel chemokine expressed in experimental inflammatory demyelinating disorders that promotes recruitment of macrophages to the central nervous system. *Am J Pathol* 154: 1181–1191

10 Glabinski AR, Tuohy VK, Ransohoff RM (1998) Expression of chemokines RANTES, MIP-1alpha and GRO-alpha correlates with inflammation in acute experimental autoimmune encephalomyelitis. *Neuroimmunomodulation* 5: 166–171

11 Glabinski AR, Tani M, Strieter RM, Tuohy VK, Ransohoff RM (1997) Synchronous synthesis of α- and β-chemokines by cells of diverse lineage in the central nervous system of mice with relapses of chronic experimental autoimmune encephalomyelitis. *Am J Pathol* 150: 617–630

12 Karpus WJ, Lukacs NW, McRae BL, Strieter RM, Kunkel SL, Miller SD (1995) An important role for the chemokine macrophage inflammatory protein-1α in the pathogenesis of the T cell-mediated autoimmune disease, experimental autoimmune encephalomyelitis. *J Immunol* 155: 5003–5010

13 Fife BT, Kennedy KJ, Paniagua MC, Lukacs NW, Kunkel SL, Luster AD, Karpus WJ (2001) CXCL10 (IFN-gamma-inducible protein-10) control of encephalitogenic CD4+ T cell accumulation in the central nervous system during experimental autoimmune encephalomyelitis. *J Immunol* 166: 7617–7624

14 Karpus WJ, Fife BT, Kennedy KJ (2003) Immunoneutralization of chemokines for the prevention and treatment of central nervous system autoimmune disease. *Methods* 29: 362–368

15 Kohler RE, Caon AC, Willenborg DO, Clark-Lewis I, McColl SR (2003) A role for macrophage inflammatory protein-3α/CC chemokine ligand 20 in immune priming during T cell-mediated inflammation of the central nervous system. *J Immunol* 170: 6298

16 Karpus WJ, Ransohoff RM (1998) Chemokine regulation of experimental autoimmune encephalomyelitis: temporal and spatial expression patterns govern disease pathogenesis. *J Immunol* 161: 2667–2671

17 Huang DR, Wang J, Kivisakk P, Rollins BJ, Ransohoff RM (2001) Absence of monocyte chemoattractant protein 1 in mice leads to decreased local macrophage recruitment and antigen-specific T helper cell type 1 immune response in experimental autoimmune encephalomyelitis. *J Exp Med* 193: 713–726

18 Klein RS, Izikson L, Means T, Gibson HD, Lin E, Sobel RA, Weiner HL, Luster AD (2004) IFN-inducible protein 10/CXC chemokine ligand 10-independent induction of experimental autoimmune encephalomyelitis. *J Immunol* 172: 550–559

19 Elhofy A, Wang J, Tani M, Fife BT, Kennedy KJ, Bennett J, Huang D, Ransohoff RM, Karpus WJ (2005) Transgenic expression of CCL2 in the central nervous system prevents experimental autoimmune encephalomyelitis. *J Leukoc Biol* 77: 229–237

20 Karpus WJ, Lukacs NW, Kennedy KJ, Smith WS, Hurst SD, Barrett TA (1997) Differential CC chemokine-induced enhancement of T helper cell cytokine production. *J Immunol* 158: 4129–4136

21 Gu L, Tseng S, Horner RM, Tam C, Loda M, Rollins BJ (2000) Control of TH2 polarization by the chemokine monocyte chemoattractant protein-1. *Nature* 404: 407–411

22 Jiang Y, Salafranca MN, Adhikari S, Xia Y, Feng L, Sonntag MK, deFiebre CM, Pennell NA, Streit WJ, Harrison JK (1998) Chemokine receptor expression in cultured glia and rat experimental allergic encephalomyelitis. *J Neuroimmunol* 86: 1–12

23 Charles PC, Weber KS, Cipriani B, Brosnan CF (1999) Cytokine, chemokine and chemokine receptor mRNA expression in different strains of normal mice: implications for establishment of a Th1/Th2 bias. *J Neuroimmunol* 100: 64–73

24 Rajan AJ, Asensio VC, Campbell IL, Brosnan CF (2000) Experimental autoimmune encephalomyelitis on the SJL mouse: effect of gamma delta T cell depletion on chemokine and chemokine receptor expression in the central nervous system. *J Immunol* 164: 2120–2130

25 Matejuk A, Vandenbark AA, Burrows GG, Bebo BF Jr., Offner H (2000) Reduced

chemokine and chemokine receptor expression in spinal cords of TCR BV8S2 transgenic mice protected against experimental autoimmune encephalomyelitis with BV8S2 protein. *J Immunol* 164: 3924–3931

26 Rottman JB, Slavin AJ, Silva R, Weiner HL, Gerard CG, Hancock WW (2000) Leukocyte recruitment during onset of experimental allergic encephalomyelitis is CCR1 dependent. *Eur J Immunol* 30: 2372–2377

27 Fife BT, Huffnagle GB, Kuziel WA, Karpus WJ (2000) CC chemokine receptor 2 is critical for induction of experimental autoimmune encephalomyelitis. *J Exp Med* 192: 899–906

28 Gao JL, Wynn TA, Chang Y, Lee EJ, Broxmeyer HE, Cooper S, Tiffany HL, Westphal H, Kwon-Chung J, Murphy PM (1997) Impaired host defense, hematopoiesis, granulomatous inflammation and type 1-type 2 cytokine balance in mice lacking CC chemokine receptor 1. *J Exp Med* 185: 1959–1968

29 Izikson L, Klein RS, Charo IF, Weiner HL, Luster AD (2000) Resistance to experimental autoimmune encephalomyelitis in mice lacking the CC chemokine receptor (CCR)2. *J Exp Med* 192: 1075–1080

30 Tran EH, Kuziel WA, Owens T (2000) Induction of experimental autoimmune encephalomyelitis in C57BL/6 mice deficient in either the chemokine macrophage inflammatory protein- 1alpha or its CCR5 receptor. *Eur J Immunol* 30: 1410–1415

31 Gaupp S, Pitt D, Kuziel WA, Cannella B, Raine CS (2003) Experimental autoimmune encephalomyelitis (EAE) in CCR2(-/-) mice: susceptibility in multiple strains. *Am J Pathol* 162: 139–150

32 Hesselgesser J, Ng HP, Liang M, Zheng W, May K, Bauman JG, Monahan S, Islam I, Wei GP, Ghannam A et al (1998) Identification and characterization of small molecule functional antagonists of the CCR1 chemokine receptor. *J Biol Chem* 273: 15687–15692

33 Liang M, Mallari C, Rosser M, Ng HP, May K, Monahan S, Bauman JG, Islam I, Ghannam A, Buckman B et al (2000) Identification and characterization of a potent, selective, and orally active antagonist of the CC chemokine receptor-1. *J Biol Chem* 275: 19000–19008

34 Delovitch TL, Singh B (1997) The nonobese diabetic mouse as a model of autoimmune diabetes: immune dysregulation gets the NOD. *Immunity* 7: 727–738

35 Christen U, Von Herrath MG (2002) Transgenic animal models for type 1 diabetes: linking a tetracycline-inducible promoter with a virus-inducible mouse model. *Transgenic Res* 11: 587–595

36 Meagher C, Sharif S, Hussain S, Cameron MJ, Arreaza GA, Delovitch TL (2003) Cytokines and chemokines in the pathogenesis of murine type 1 diabetes. *Adv Exp Med Biol* 520: 133–158

37 Arimilli S, Ferlin W, Solvason N, Deshpande S, Howard M, Mocci S (2000) Chemokines in autoimmune diseases. *Immunol Rev* 177: 43–51

38 Cardozo AK, Proost P, Gysemans C, Chen MC, Mathieu C, Eizirik DL (2003) IL-1beta and IFN-gamma induce the expression of diverse chemokines and IL-15 in human and

rat pancreatic islet cells, and in islets from pre-diabetic NOD mice. *Diabetologia* 46: 255–266

39 Bradley LM, Asensio VC, Schioetz LK, Harbertson J, Krahl T, Patstone G, Woolf N, Campbell IL, Sarvetnick N (1999) Islet-specific Th1, but not Th2, cells secrete multiple chemokines and promote rapid induction of autoimmune diabetes. *J Immunol* 162: 2511–2520

40 Cameron MJ, Arreaza GA, Grattan M, Meagher C, Sharif S, Burdick MD, Strieter RM, Cook DN, Delovitch TL (2000) Differential expression of CC chemokines and the CCR5 receptor in the pancreas is associated with progression to type I diabetes. *J Immunol* 165: 1102–1110

41 Christen U, McGavern DB, Luster AD, Von Herrath MG, Oldstone MB (2003) Among CXCR3 chemokines, IFN-gamma-inducible protein of 10 kDa (CXC chemokine ligand (CXCL) 10) but not monokine induced by IFN-gamma (CXCL9) imprints a pattern for the subsequent development of autoimmune disease. *J Immunol* 171: 6838–6845

42 Kim SH, Cleary MM, Fox HS, Chantry D, Sarvetnick N (2002) CCR4-bearing T cells participate in autoimmune diabetes. *J Clin Invest* 110: 1675–1686

43 Camacho SA, Heath WR, Carbone FR, Sarvetnick N, LeBon A, Karlsson L, Peterson PA, Webb SR (2001) A key role for ICAM-1 in generating effector cells mediating inflammatory responses. *Nat Immunol* 2: 523–529

44 Savinov AY, Wong FS, Stonebraker AC, Chervonsky AV (2003) Presentation of antigen by endothelial cells and chemoattraction are required for homing of insulin-specific CD8+ T cells. *J Exp Med* 197: 643–656

45 Grewal IS, Rutledge BJ, Fiorillo JA, Gu L, Gladue RP, Flavell RA, Rollins BJ (1997) Transgenic monocyte chemoattractant protein-1 (MCP-1) in pancreatic islets produces monocyte-rich insulitis without diabetes. Abrogation by a second transgene expressing systemic MCP-1. *J Immunol* 159: 401–408

46 Morimoto J, Yoneyama H, Shimada A, Shigihara T, Yamada S, Oikawa Y, Matsushima K, Saruta T, Narumi S (2004) CXC chemokine ligand 10 neutralization suppresses the occurrence of diabetes in nonobese diabetic mice through enhanced beta cell proliferation without affecting insulitis. *J Immunol* 173: 7017–7024

47 Frigerio S, Junt T, Lu B, Gerard C, Zumsteg U, Hollander GA, Piali L (2002) Beta cells are responsible for CXCR3-mediated T-cell infiltration in insulitis. *Nat Med* 8: 1414–1420

48 Ransohoff RM, Bacon KB (2000) Chemokine receptor antagonism as a new therapy for multiple sclerosis. *Expert Opin Investig Drugs* 9: 1079–1097

49 Liang M, Mallari C, Rosser M, Ng HP, May K, Monahan S, Bauman JG, Islam I, Ghannam A, Buckman B et al (2000) Identification and characterization of a potent, selective, and orally active antagonist of the CC chemokine receptor-1. *J Biol Chem* 275: 19000–19008

Chemokines in allergic responses: eosinophils, basophils, mast cells

Zamaneh Mikhak and Andrew D. Luster

Center for Immunology and Inflammatory Diseases, Division of Rheumatology, Allergy and Immunology, Massachusetts General Hospital, Harvard Medical School, 149 Thirteenth Street, Room 8301, Charlestown, MA 02129, USA

Introduction

Eosinophils, basophils and mast cells play key roles in the allergic response. These cells are cellular members of the innate immune system and contain granules with a variety of potent biological mediators. Mast cells are tissue bound and positioned near epithelial surfaces and as such can respond quickly to tissue injury, parasites and allergens by releasing the content of their granules. Eosinophils and basophils circulate within the blood stream and traffic to sites of tissue damage and parasite/allergen exposure. They too release potent biological mediators upon activation. The activation of mast cells, eosinophils and basophils and the subsequent release of their granules lead to many of the phenotypic features observed in the allergic response, such as vasodilatation and tissue edema.

Participation of eosinophils, basophils and mast cells in the allergic response is not limited to the release of their pre-formed biological mediators. Eosinophils, basophils and mast cells contribute to the allergic response in at least three additional ways: they provide a link between the innate and adaptive immune response during the initial phase of allergic inflammation, they skew the adaptive immune response to new antigens towards a Th2 phenotype and they amplify the allergic response during viral infections through the function of their chemokines.

In 1989, the presence of heat labile eosinophil chemotactic activities in the serum of atopic patients during seasonal pollen exposure was first described [1]. Today, chemokines and their G-protein coupled receptors have emerged as key controllers in the movement of eosinophils, basophils and mast cells to sites of allergic inflammation with significant implications for therapy. This chapter focuses on the role of eosinophils, basophils and mast cells in allergic responses and suggests that these cells participate in many aspects of the allergic response through the function of their chemokines and chemokine receptors.

Chemokine Biology – Basic Research and Clinical Application, Volume II
edited by Kuldeep Neote, Gordon L. Letts and Bernhard Moser

Eosinophils, basophils and mast cells play a key role in allergic inflammation through the function of their chemokines and chemokine receptors

Antigen presenting cells become activated upon exposure to antigens/allergens and release cytokines. The cytokine milieu of the tissue, in turn, leads to the expression of distinct patterns of chemokines by resident cells and inflammatory cells. This local production of chemokines is crucial in attracting immune cells that bear corresponding chemokine receptors to sites of antigen entry and deposition. Once immune cells arrive in peripheral tissues, they produce inflammatory mediators and additional chemokines that augment the inflammatory response. The generation of the allergic response requires that the chemokine milieu of the peripheral tissue attract and activate eosinophils, basophils and mast cells. These cells participate in the allergic response by both releasing chemokines and responding to local chemokines through the function of their chemokine receptors.

Local production of chemokines is a significant finding in many allergic disorders. For example, bronchial biopsies and bronchoalveolar lavage (BAL) from atopic asthmatic patients after allergen challenge show an increase in CCL11 gene expression and protein levels as well as an influx of eosinophils at 4 h post challenge [2]. Moreover, in a study of asthmatic children, CCL5, CCL7, CCL13, CCL11 and CCL24 levels were significantly increased in the BAL fluid [3]. In general, CCL2, CCL5, CCL7, CCL8, CCL11, CCL13, CCL17, CCL22, CCL24 and CCL26 are highly expressed in peripheral tissues in allergic inflammation and the receptors for these chemokines, CCR2, CCR3 and CCR4 are expressed on eosinophils [4–6] while mast cells [7] and basophils express CCR2 and CCR3 [8–9].

Antigen presenting cells capture antigens in peripheral tissues and traffic to draining lymph nodes where they present antigens to naive T cells (Fig. 1). Activation and differentiation of naive T cells into Th2 cells is accompanied by expression of a distinct subset of chemokine receptors, CCR4 and CCR8, on Th2 cells. These chemokine receptors enable Th2 cells to traffic to sites of allergic inflammation by sensing the concentration gradient established at allergic inflammatory sites for their ligands: CCL17, CCL22 and CCL1. Once in peripheral tissues, Th2 cells release IL4, IL5 and IL13. While IL5 expression leads to release of eosinophils from the bone marrow, IL4 and IL13 work on resident cells such as macrophages, epithelial cells and endothelial cells by activating transcription factor STAT6 and upregulating the production of chemokines, CCL11, CCL24, CCL26, CCL1, CCL17 and CCL22. This process helps further establish the concentration gradient for chemokines at allergic inflammatory sites and forms the basis for STAT6 dependency of Th2-cell trafficking [10].

Meanwhile, circulating eosinophils, basophils and mast cells express chemokine receptors that enable them to respond to the chemokines established at allergic inflammatory sites. In particular, all three cell types express CCR3, which equips

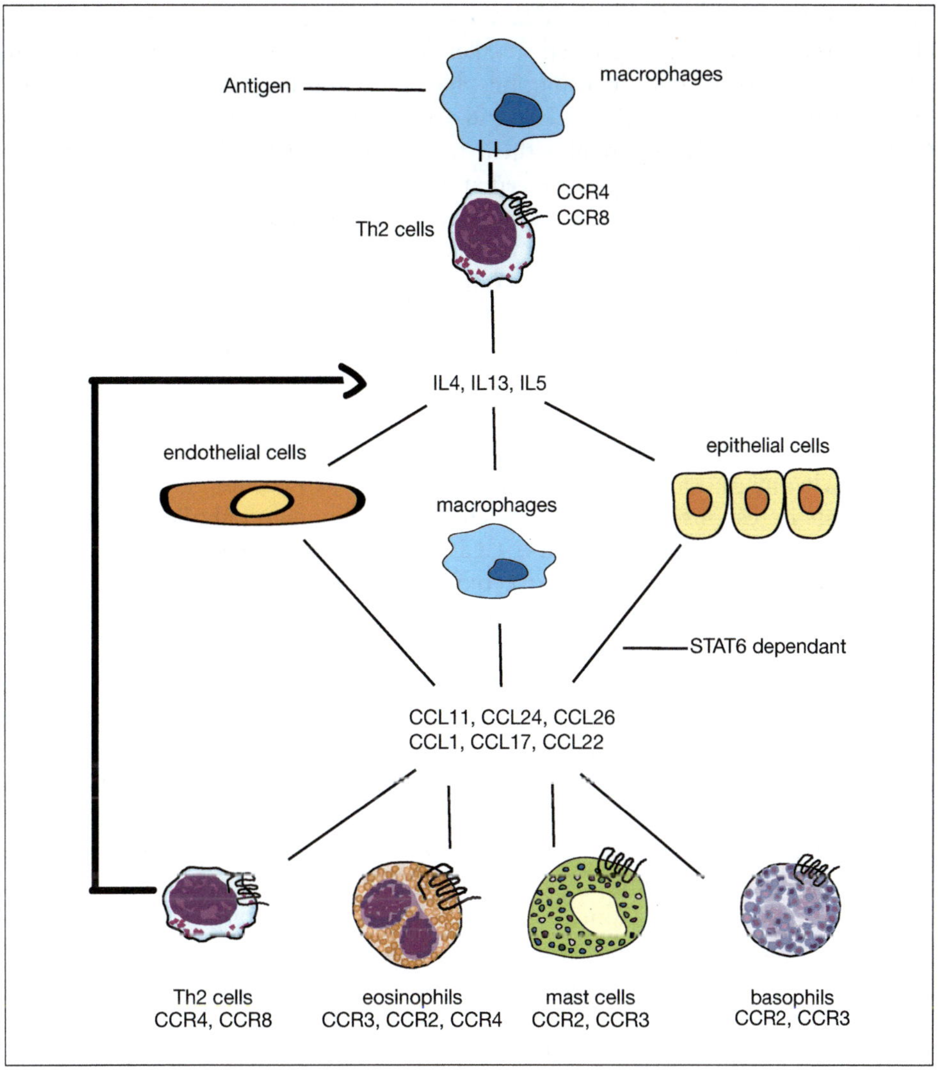

Figure 1
Eosinophils, basophils and mast cells play a key role in allergic inflammation through the function of their chemokines and chemokine receptors: Antigen-presenting cells present antigen to naive T cells, which differentiate to Th2 cells and release IL4, IL5 and IL13. These cytokines activate resident cells, in a STAT6 dependant manner, to release of CCL11, CCL24, CCL26, CCL1, CCL17 and CCL22. Eosinophils, basophils and mast cells as well as Th2 cells respond to the chemokines generated by resident cells through the function of their chemokine receptors, traffic to sites of allergic inflammation, and further release IL4, IL5 and IL13 to augment the allergic response.

them to traffic towards CCL11, CCL24, CCL26, CCL5, CCL7, CCL8 and CCL13, chemokines abundantly expressed at allergic inflammatory sites. In addition to CCR3, CCR2 expression on eosinophils, basophils and mast cells leads to chemotaxis towards CCL2, CCL7, CCL8, and CCL13 [11]. Finally, the expression of CCR4 on eosinophils [4] enables them to respond to CCL17 and CCL22, two chemokines characteristically expressed at allergic inflammatory sites.

The interaction of chemokines generated at sites of allergic inflammation with chemokine receptors on eosinophils results in upregulation of adhesion molecules and enhances trafficking. For instance, CCL11 upregulates the level of CD11b/CD18 on eosinophils in the presence of IL5 and increases the adhesion of eosinophils to fibronectin as well as ICAM on activated normal human bronchial epithelial cell monolayers [12, 13]. Similarly, eosinophil migration across activated human pulmonary microvascular endothelial cells (HPMEC) that express VCAM-1 depends on both alpha4 integrin and CCR3 [14].

Following migration into peripheral tissues, activated eosinophils and basophils degranulate in response to many of the chemokines abundant at allergic inflammatory sites, such as CCL11, CCL24, CCL13 and CCL2, CCL7, CCL8, CCL13, respectively [11] and release reactive oxygen species and cysteinyl leukotrienes. Eosinophils are not only a main source of LTC4, LTD4 and LTE4, but also are able to respond to these mediators through their cysteinyl leukotrienes receptors, CysLT1 and CysLT2 [15]. Cysteinyl leukotrienes are potent mediators that prolong eosinophil survival, contribute to eosinophil chemotaxis, augment eosinophil adhesion to airway epithelium through interaction of beta integrins with ICAM on epithelial cells, initiate the respiratory burst in eosinophils and cause bronchial smooth muscle cell spasm [16]. Activated eosinophils also generate a large array of chemokines, including CCL2, CCL3, CCL4, CCL5, and CCL11, which further amplify the allergic response [17–20]. Activated basophils release CCL3 [21], leukotrienes and prostoglandins, which lead to the phenotypic features of vasodilatation and increased vascular permeability seen in allergic responses [22].

The degranulation of eosinophils, basophils and mast cells and the release of their active metabolites further augment the allergic cascade by amplifying the chemokine milieu at allergic inflammatory sites leading to increased recruitment of not only Th2 cells but also eosinophils, basophils and mast cells.

Eosinophils, basophils and mast cells link innate and adaptive immunity during allergic inflammation: Role of chemokines and chemokine receptors

During the initial phase of the allergic response, antigen-presenting cells directly interact with foreign antigens and release cytokines that mediate the production of chemokines from resident cells at inflammatory sites. This initial local production of chemokines is important for attracting the first wave of Th2 cells to sites of allergic inflammation. Once Th2 cells are recruited to inflammatory sites, they efficiently produce more cytokines and chemokines to augment the allergic response. In addition to antigen-presenting cells, eosinophils, basophils and mast cells, as members of the innate immune response, contribute to this initial local chemokine milieu that attracts the first wave of Th2 cells. In this way, eosinophils, basophils and mast cells link the innate immune response to adaptive immunity during allergic inflammation.

The initial local production of cytokines and chemokines is a process that is mediated through the innate immune system. For example, release of IL-1β and TNF induces epithelial cells to express CCL13 [23] while IL-1β-stimulated A549 airway epithelial cells produce CCL11 [24]. Furthermore, smooth muscle cells express CCL11, CCL2, CCL8, CCL7 and CCL5 after exposure to IL-1 and TNF [25, 26]. Moreover, platelet activation during allergen challenge leads to the release of many platelet products, including CCL5 [27] while TGF-beta induces the production of CCL11 in fibroblasts [28].

In addition to antigen presenting cells, resident mast cells play an important role in directing the peripheral tissue towards chemokine production in early phases of the allergic response. Resident mast cells are positioned to respond to new antigens through the function of their complement receptors and toll like receptors, which are activated when antigens cause local damage or contain recognizable repeatable patterns. For example, house dust mites and fungi have proteases that cleave complement [29, 30] and cause damage at the site of tissue entry. C3a and C5a attract and activate resident mast cells that bear C3a and C5a receptors [31]. In addition to releasing histamine, tryptase, chymase, proteases and arachadonic acid metabolites, activated mast cells release a large number of chemokines, such as CCL1, CCL2, CCL3, CCL4, CCL5, CCL8, CCL17 and CCL22 [7, 32–33].

Resident mast cells also bear toll-like receptors that become activated by their interaction with LPS, viral particles, and peptidoglycan. For example, in a study of bone marrow derived mast cells from wild type and TLR2 and TLR4 deficient mice, mast cells were stimulated with peptidoglycan and LPS and patterns of cytokine and chemokine secretion were assessed by ELISA. TLR2 activation of bone marrow derived mast cells led to secretion of IL4, IL5 and IL13 while TLR4 activation of these cells resulted in IL13 secretion [34]. Furthermore, LPS activation of TLR4 in mast cells has been associated with the release of Th2 type chemokines, CCL1, CCL5, and CCL8 [32]. These reports are consistent with findings that low levels of LPS signaling through TLR4, in a TNF dependent fashion, are required for Th2 responses to inhaled antigens [35]. Similar to mast cells, basophils bear TLR2 and respond to peptidoglycan by secreting IL4 and IL13 [36].

Eosinophils also play an important role in promoting cytokine and chemokine production in the tissue during the innate immune response. A study of the immune response to *N. brasiliensis* infection in $Rag1^{-/-}$ and wild type mice revealed that in the absence of T cells and B cells in Rag1$^{-/-}$ mice, eosinophils are able to respond to *N. brasiliensis*, accumulate in the tissue and produce IL-4 [37]. Cytokines such as IL-4 and IL-13, in turn, lead to the production of many chemokines, including CCL17 and CCL22 [11]. Th2 cells respond to CCL1, CCL17 and CCL22 secreted by mast cells through CCR8 and CCR4 and home to the site of inflammation. Arrival of Th2 cells to sites of allergic inflammation links the initial innate immune response to allergens to the adaptive immune response in which antigen-specific Th2 cells specifically recognize allergens. Upon arrival, Th2 cells secrete IL-4 and IL-13 in the peripheral tissues. These Th2 type cytokines can then efficiently further drive the production of many chemokines, such as CCL11, CCL24, CCL26, CCL7, CCL13, CCL17 and CCL22 [11] (Fig. 2).

Eosinophils, basophils and mast cells skew the adaptive immune response to new antigens towards Th2 inflammation: Role of chemokines and chemokine receptors

The allergic response is characterized by the generation of antigen-specific Th2 cells, which bind their cognate antigen and release potent cytokines such as IL-4 and IL-13. These cytokines then direct the inflammatory process by regulating the chemokine milieu of the peripheral tissue, controlling the type of cells recruited to the inflammatory site, directing antibody class switching and influencing resident cells to, for example, undergo mucous hypersecretion or smooth muscle contraction. While Th2 cells are a significant source of IL4, their generation at the same time requires IL-4 signaling through the STAT6 pathway. The initial differentiation of T cells towards a Th2 phenotype, therefore, requires a source of IL-4 independent of Th2 cells.

Activated mast cells, eosinophils and basophils secrete IL-4 and can provide the original IL-4 required to initiate Th2 differentiation. Interestingly, while T cell differentiation into IL-4-producing Th2 cells requires IL-4 signaling through Stat6, mast cells, eosinophils and basophils produce IL-4 in a mechanism independent of IL-4 and Stat6 [38, 39]. Once Th2 cells are generated, they secrete additional IL4 for further Th2 differentiation.

IL-4 production by mast cells requires Stem Cell Factor (SCF), an important regulator, which is produced by eosinophils, mast cells, epithelial cells and macrophages. The receptor for SCF, c-kit, is expressed on eosinophils and mast cells and SCF promotes mast cell and eosinophil maturation, activation, recruitment and survival [40]. Bone marrow derived mast cells cultured with primary pulmonary fibroblast cell lines, which constitutively produce SCF in its trans-membrane form, have been found to upregulate CCL11 production, a process that depends on cell contact and is inhibited by anti-SCF [41]. SCF upregulates the eosinophil release of CCL4, CCL5, CCL17 and CCL22 [20], the mast cell release of CCL2, CCL5,

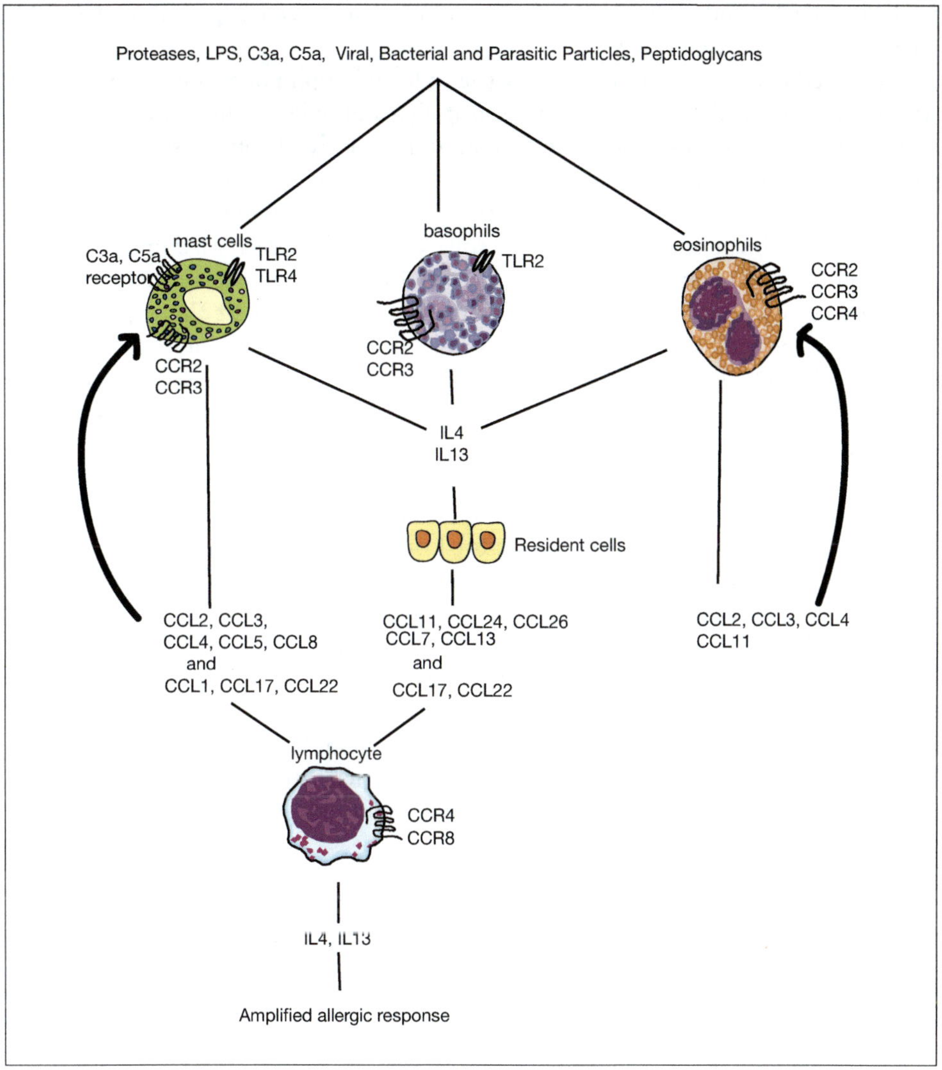

Figure 2

Eosinophils, basophils and mast cells link innate and adaptive immunity during allergic inflammation: Proteases, lipopolysaccharide (LPS), complement fragments, viral, bacterial and parasitic particles and peptidoglycan activate eosinophils, basophils and mast cells to secrete IL4 and IL13 as well as a number of chemokines. CCL2, CCL3, CCL4, CCL5, CCL7, CCL8, CCL13, CCL11, CCL24 and CCL26 attract more eosinophils, basophils and mast cells while CCL1, CCL17 and CCL22 lead to the recruitment of Th2 cells that bear CCR8 and CCR4. Upon arrival, Th2 cells secrete IL4, IL5 and IL13, which further augment the allergic response.

CCL11, CCL17 and CCL22 [7, 40, 41] and the mast cell expression of CCR1, CCR2, CCR3 and CCR5 [7].

Activated mast cells migrate from tissue to local lymph nodes during the process of sensitization in delayed type hypersensitivity and subsequently promote T cell recruitment into the lymph nodes [42] (Fig. 3). In a study of contact sensitivity, dinitrofluorobenzene *versus* vehicle was painted on the shaved abdomen or footpad of mice on two consecutive days and the skin and draining lymph nodes were harvested at different time points. With time, density of mast cells decreased in the skin but increased in the draining lymph nodes. *In situ* hybridization showed that infiltrating mast cells were the predominant source of CCL4 and that the expression of CCL4 corresponded with the influx of $CD4^+$ and $CD8^+$ T cells into the draining lymph node [42, 43]. Neutralization of this chemokine during the sensitization phase inhibited $CD4^+$ and $CD8^+$ T cell trafficking into the draining lymph node and decreased delayed type hypersensitivity responses [43].

Mast cells and basophils have also been shown to express CD40 ligand and induce IgE production in B cells *in vitro* [44, 45]. Once the allergic response is established, IgE occupies the Fc epsilon receptor I on mast cells. Upon reencounter with the allergen, the reaction of the mast cell to the allergen is amplified as the cross linking of IgE by allergen upregulates CCL1, CCL2, CCL3, CCL4, CCL7, CCL17, CCL22 and XCL1 production [32, 46–49] as well as CCR1, CCR2, CCR3 and CCR5 expression [7]. Finally, the balance between Th1 and Th2 inflammation is at least partly controlled by the cytokine milieu at the time of T cell differentiation and histamine release by mast cells has been shown to modulate cytokine production in favor of Th2 inflammation. Specifically, histamine has been shown to enhance the secretion of IL-5, IL-10 and IL-13 while suppressing the production of interferon gamma [50].

Eosinophils basophils and mast cells amplify the allergic response through the function of their chemokines

The benefits of the inflammatory response generated by eosinophils, basophils and mast cells in fighting parasites are easy to appreciate. However, it is not clear why atopic individuals respond to antigens that are harmless to others with allergic inflammation. It has been postulated that allergic mechanisms serve a beneficial end and it is the over activation or inappropriate activation of allergic processes that leads to the development of allergic disorders. For example, aspects of the allergic inflammatory cascade may help fight infections with viruses and this, in part, may explain why atopic disorders tend to worsen during viral infections.

Viral infections contribute to the allergic response at least partly by increasing the recruitment of antigen-specific and antigen non-specific Th2 cells into inflammatory sites [51]. The influx of Th2 cells increases the level of Th2 cytokines, such

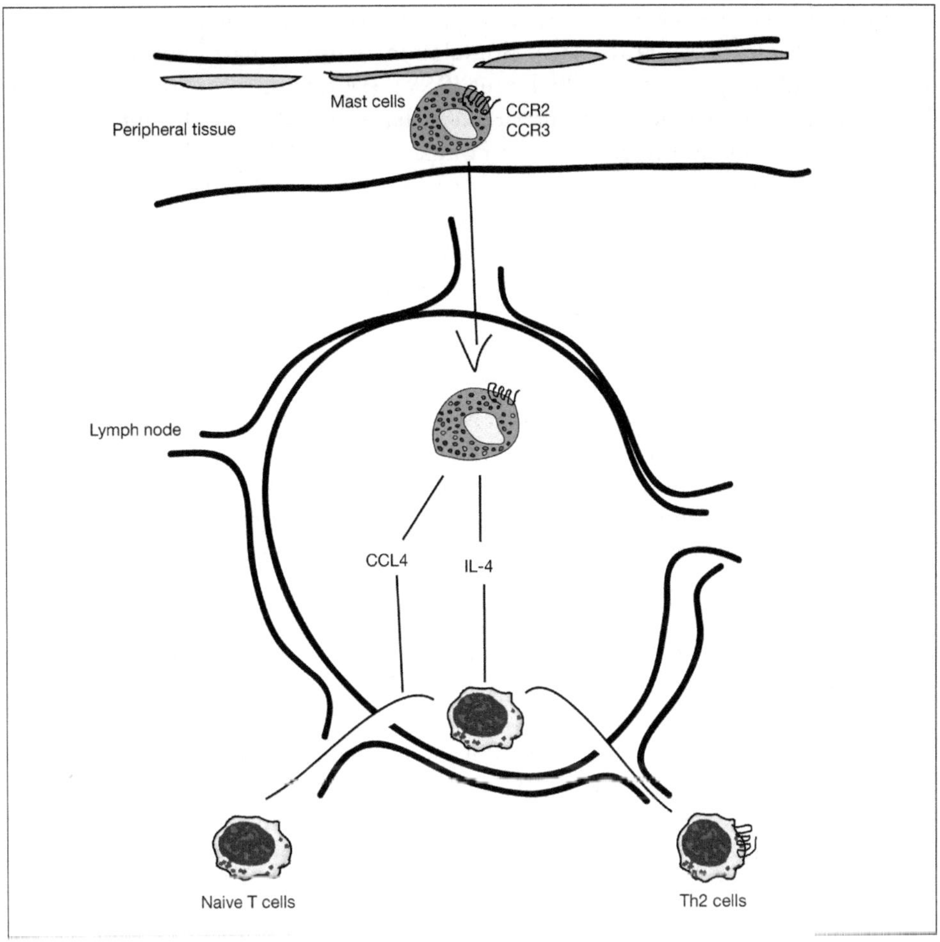

Figure 3
Mast cells migrate to draining lymph nodes in response to CCL4. Once in the lymph node, mast cells are the predominant source of CCL4. Levels of CCL4 correlate with the influx of T cells into the lymph node. CCL4 neutralization results in decreased T cell trafficking into the lymph node and attenuation of the allergic response.

as IL-4 and IL-13, in the tissue and upregulates Th2-associated chemokines accordingly. The enhanced level of Th2-associated chemokines recruits more inflammatory cells such as Th2 cells, eosinophils, basophils and mast cells and increases the allergic inflammatory response.

Murine fetal skin-derived mast cells have been shown to express TLR3 and TLR7 and produce CCL3 and CCL5 in response to poly (I:C) and R-848 [52]. Furthermore, human peripheral blood-derived cultured mast cells as well as human mast cell lines express TLR3 [53]. Stimulation of these mast cells with double stranded RNA, respiratory syncytial virus (RSV), UV-treated influenza virus or type I reovirus leads to production of interferon alpha. Antibody to TLR3 inhibits interferon alpha production in response to double stranded RNA and mast cells from TLR3 knock out mice show attenuated responses to double stranded RNA [53]. Eosinophils express TLR7 and generate superoxide in response to the ligand for TLR7, R-848 [54]. Moreover, infection with RSV increases the production of CCL3 and CCL5 in airway epithelial cells and eosinophils [55] while activated mast cells and basophils release CCL3 [21, 56].

The production of CCL3 during viral infections results in more effective cell mediated immune responses as CCL3 contributes to $CD11c^{+}CD11b^{+}CD8alpha^{-}$ dendritic cell maturation and activation as well as dendritic cell migration into the draining lymph nodes [57]. In addition, CCL3 promotes $CD8^{+}$ effector T cell differentiation and recruitment. CCL3 deficient mice infected in the central nervous system (CNS) with a mouse hepatitis virus show decreased macrophage and $CD8^{+}$ T cell infiltration into the CNS, decreased viral clearing and decreased $CD8^{+}$ T cell differentiation into effector cells as well as decreased cytolytic activity [58]. CCL3 and CCL5 have also been shown to enhance natural killer (NK) cell chemotaxis and increase NK cell granule exocytosis and cytolytic activity [59]. Moreover, these chemokines increase macrophage chemotaxis [58] and macrophage activation by upregulating CD40, IL-12 and TNF production [60].

While many chemokines play an important role in enhancing cell mediated immunity against viruses, they also directly contribute to the pathophysiology observed in allergic disorders such as asthma. For example, CCL5 has been linked to airway hyperreactivity in a mouse model of RSV infection, where neutralization of CCL5 with anti-CCL5 antibody results in a significant decrease in airway hyperreactivity [61]. Furthermore, the increased levels of CCL5 lead to enhanced chemotaxis of eosinophils, basophils and mast cells and augmentation of the allergic response. The allergic phenotype becomes amplified during viral infections as many chemokines generated by eosinophils and mast cells help enhance cell-mediated immunity against viral pathogens.

In conclusion, eosinophils, basophils and mast cells participate in the allergic response in a variety of ways through the function of their chemokines and chemokine receptors. Eosinophils, basophils and mast cells contribute to the generation of chemokines at early stages of allergic inflammation, thus creating the chemokine milieu needed to recruit the first wave of Th2 cells into allergic inflammatory sites. In doing so, they help link the innate immune response to adaptive immunity. Once Th2 cells arrive at sites of allergic inflammation, they generate a large number of chemokines, which further attract and activate eosinophils, basophils and mast cells through interaction with the chemokine receptors on these cells. Eosinophils, basophils and mast cells can secrete IL-4, independent of T cell derived IL-4, and thus are capable of driving the early stage of T cell differentiation towards a Th2 phenotype. Finally, many chemokines secreted by eosinophils, basophils and mast cells potentiate cell-mediated immunity against viral pathogens and amplify the allergic response during viral infections.

References

1 Hakansson L, Rak S, Dahl R, Venge P (1989) The formation of eosinophil and neutrophil chemotactic activity during a pollen season and after allergen challenge. *J Allergy Clin Immunol* 83(5): 933–939

2 Brown JR, Kleimberg J, Marini M, Sun G, Bellini A, Mattoli S (1998) Kinetics of eotaxin expression and its relationship to eosinophil accumulation and activation in bronchial biopsies and bronchoalveolar lavage (BAL) of asthmatic patients after allergen inhalation. *Clin Exp Immunol* 114(2): 137–146

3 Rojas-Ramos E, Avalos AF, Perez-Fernandez L, Cuevas-Schacht F, Valencia-Maqueda E, Teran LM (2003) Role of the chemokines RANTES, monocyte chemotactic proteins-3 and -4, and eotaxins-1 and -2 in childhood asthma. *Eur Respir J* 22(2): 310–316

4 Liu LY, Jarjour NN, Busse WW, Kelly EA (2003) Chemokine receptor expression on human eosinophils from peripheral blood and bronchoalveolar lavage fluid after segmental antigen challenge. *J Allergy Clin Immunol* 112(3): 556–562

5 Borchers MT, Ansay T, DeSalle R, Daugherty BL, Shen H, Metzger M, Lee NA, Lee JJ (2002) *In vitro* assessment of chemokine receptor-ligand interactions mediating mouse eosinophil migration. *J Leukoc Biol* 71(6): 1033–1041

6 Dunzendorfer S, Kaneider NC, Kaser A, Woell E, Frade JM, Mellado M, Martinez-Alonso C, Wiedermann CJ (2001) Functional expression of chemokine receptor 2 by normal human eosinophils. *J Allergy Clin Immunol* 108(4): 581–587

7 Oliveira SH, Lukacs NW (2001) Stem cell factor and igE-stimulated murine mast cells produce chemokines (CCL2, CCL17, CCL22) and express chemokine receptors. *Inflamm Res* 50(3): 168–174

8 Uguccioni M, Mackay CR, Ochensberger B, Loetscher P, Rhis S, LaRosa GJ, Rao P, Ponath PD, Baggiolini M, Dahinden CA (1997) High expression of the chemokine receptor CCR3 in human blood basophils. Role in activation by eotaxin, MCP-4, and other chemokines. *J Clin Invest* 100(5): 1137–1143

9 Ochensberger B, Tassera L, Bifrare D, Rihs S, Dahinden CA (1999) Regulation of cytokine expression and leukotriene formation in human basophils by growth factors, chemokines and chemotactic agonists. *Eur J Immunol* 29(1): 11–22

10 Mathew A, MacLean JA, DeHaan E, Tager AM, Green FH, Luster AD (2001) Signal transducer and activator of transcription 6 controls chemokine production and T helper cell type 2 cell trafficking in allergic pulmonary inflammation. *J Exp Med* 193(9): 1087–1096

11 Luster AD (2001) Antichemokine immunotherapy for allergic diseases. *Curr Opin Allergy Clin Immunol* 1(6): 561–567

12 Lundahl J, Moshfegh A, Gronneberg R, Hallden G (1998) Eotaxin increases the expression of CD11b/CD18 and adhesion properties in IL5, but not fMLP-prestimulated human peripheral blood eosinophils. *Inflammation* 22(2): 123–135

13 Burke-Gaffney A, Hellewell PG (1998) A CD18/ICAM-1-dependent pathway mediates eosinophil adhesion to human bronchial epithelial cells. *Am J Respir Cell Mol Biol* 19(3): 408–418

14 Nagata M, Yamamoto H, Tabe K, Sakamoto Y (2001) Eosinophil transmigration across VCAM-1-expressing endothelial cells is upregulated by antigen-stimulated mononuclear cells. *Int Arch Allergy Immunol* 125 (Suppl 1): 7–11

15 Bandeira-Melo C, Weller PF (2003) Eosinophils and cysteinyl leukotrienes. *Prostaglandins Leukot Essent Fatty Acids* 69(2–3): 135–143

16 Nagata M, Saito K (2003) The roles of cysteinyl leukotrienes in eosinophilic inflammation of asthmatic airways. *Int Arch Allergy Immunol* 131 (Suppl 1): 7–10

17 Costa JJ, Matossian K, Resnick MB, Beil WJ, Wong DT, Gordon JR, Dvorak AM, Weller PF, Galli SJ (1993) Human eosinophils can express the cytokines tumor necrosis factor-alpha and macrophage inflammatory protein-1 alpha. *J Clin Invest* 91(6): 2673–2684

18 Nakajima T, Yamada H, Iikura M, Miyamasu M, Izumi S, Shida H, Ohta K, Imai T, Yoshie O, Mochizuki M et al (1998) Intracellular localization and release of eotaxin from normal eosinophils. *FEBS Lett* 434(3): 226–230

19 Izumi S, Hirai K, Miyamasu M, Takahashi Y, Misaki Y, Takaishi T, Morita Y, Matsushima K, Ida N, Nakamura H et al (1997) Expression and regulation of monocyte chemoattractant protein-1 by human eosinophils. *Eur J Immunol* 27(4): 816–824

20 Oliveira SH, Taub DD, Nagel J, Smith R, Hogaboam CM, Berlin A, Lukacs NW (2002) Stem cell factor induces eosinophil activation and degranulation: mediator release and gene array analysis. *Blood* 100(13): 4291–4297

21 Li H, Sim TC, Grant JA, Alam R (1996) The production of macrophage inflammatory protein-1 alpha by human basophils. *J Immunol* 157(3): 1207–1212

22 Holgate ST (2000) The role of mast cells and basophils in inflammation. *Clin Exp Allergy* 30 (Suppl 1): 28–32

23 Lamkhioued B, Garcia-Zepeda EA, Abi-Younes S, Nakamura H, Jedrzkiewicz S, Wagner L, Renzi PM, Allakhverdi Z, Lilly C, Hamid Q et al (2000) Monocyte chemoattractant protein (MCP)-4 expression in the airways of patients with asthma. Induction in epithelial cells and mononuclear cells by proinflammatory cytokines. *Am J Respir Crit Care Med* 162(2 Pt 1): 723–732

24 Jedrzkiewicz S, Nakamura H, Silverman ES, Luster AD, Mansharamani N, In KH, Tamura G, Lilly CM (2000) IL-1beta induces eotaxin gene transcription in A549 airway epithelial cells through NF-kappaB. *Am J Physiol Lung Cell Mol Physiol* 279(6): L1058–1065

25 Singer CA, Salinthone S, Baker KJ, Gerthoffer WT (2004) Synthesis of immune modulators by smooth muscles. *Bioessays* 26(6): 646–655

26 Ghaffar O, Hamid Q, Renzi PM, Allakhverdi Z, Molet S, Hogg JC, Shore SA, Luster AD, Lamkhioued B (1999) Constitutive and cytokine-stimulated expression of eotaxin by human airway smooth muscle cells. *Am J Respir Crit Care Med* 159(6): 1933–1942

27 Kameyoshi Y, Dorschner A, Mallet AI, Christophers E, Schroder JM (1992) Cytokine RANTES released by thrombin-stimulated platelets is a potent attractant for human eosinophils. *J Exp Med* 176(2): 587–592

28 Wenzel SE, Trudeau JB, Barnes S, Zhou X, Cundall M, Westcott JY, McCord K, Chu HW (2002) TGF-beta and IL-13 synergistically increase eotaxin-1 production in human airway fibroblasts. *J Immunol* 169(8): 4613–4619

29 Maruo K, Akaike T, Ono T, Okamoto T, Maeda H (1997) Generation of anaphylatoxins through proteolytic processing of C3 and C5 by house dust mite protease. *J Allergy Clin Immunol* 100(2): 253–260

30 Washburn RG, Bryant-Varela BJ, Julian NC, Bennett JE (1991) Differences in Cryptococcus neoformans capsular polysaccharide structure influence assembly of alternative complement pathway C3 convertase on fungal surfaces. *Mol Immunol* 28(4–5): 465–470

31 Nilsson G, Johnell M, Hammer CH, Tiffany HL, Nilsson K, Metcalfe DD, Siegbahn A, Murphy PM (1996) C3a and C5a are chemotaxins for human mast cells and act through distinct receptors via a pertussis toxin-sensitive signal transduction pathway. *J Immunol* 157(4): 1693–1698

32 Okumura S, Kashiwakura J, Tomita H, Matsumoto K, Nakajima T, Saito H, Okayama Y (2003) Identification of specific gene expression profiles in human mast cells mediated by Toll-like receptor 4 and FcepsilonRI. *Blood* 102(7): 2547–2554

33 Selvan RS, Butterfield JH, Krangel MS (1994) Expression of multiple chemokine genes by a human mast cell leukemia. *J Biol Chem* 269(19): 13893–13898

34 Supajatura V, Ushio H, Nakao A, Akira S, Okumura K, Ra C, Ogawa H (2002) Differential responses of mast cell Toll-like receptors 2 and 4 in allergy and innate immunity. *J Clin Invest* 109(10):1351–1359

35 Eisenbarth SC, Piggott DA, Huleatt JW, Visintin I, Herrick CA, Bottomly K (2002) Lipopolysaccharide-enhanced, toll-like receptor 4-dependent T helper cell type 2 responses to inhaled antigen. *J Exp Med* 196(12): 1645–1651

36 Bieneman AP, Chichester KL, Chen YH, Schroeder JT (2005) Toll-like receptor 2 ligands activate human basophils for both IgE-dependent and IgE-independent secretion. *J Allergy Clin Immunol* 115(2): 295–301

37 Shinkai K, Mohrs M, Locksley RM (2002) Helper T cells regulate type-2 innate immunity *in vivo*. *Nature* 420: 825–829

38 Sherman MA, Secor VH, Lee SK, Lopez RD, Brown MA (1999) STAT6-independent production of IL-4 by mast cells. *Eur J Immunol* 29(4): 1235–1242

39 Voehringer D, Shinkai K, Locksley RM (2004) Type 2 immunity reflects orchestrated recruitment of cells committed to IL-4 production. *Immunity* 20(3): 267–277

40 Oliveira SH, Lukacs NW (2003) Stem cell factor: a hemopoietic cytokine with important targets in asthma. *Curr Drug Targets Inflamm Allergy* 2(4): 313–318

41 Hogaboam C, Kunkel SL, Strieter RM, Taub DD, Lincoln P, Standiford TJ, Lukacs NW (1998) Novel role of transmembrane SCF for mast cell activation and eotaxin production in mast cell-fibroblast interactions. *J Immunol* 160(12): 6166–6171

42 Wang HW, Tedla N, Lloyd AR, Wakefield D, McNeil PH (1998) Mast cell activation and migration to lymph nodes during induction of an immune response in mice. *J Clin Invest* 102(8): 1617–1626

43 Tedla N, Wang HW, McNeil HP, Di Girolamo N, Hampartzoumian T, Wakefield D, Lloyd A (1998) Regulation of T lymphocyte trafficking into lymph nodes during an immune response by the chemokines macrophage inflammatory protein (MIP)-1 alpha and MIP-1 beta. *J Immunol* 161(10): 5663–5672

44 Gauchat JF, Henchoz S, Mazzei G, Aubry JP, Brunner T, Blasey H, Life P, Talabot D, Flores-Romo L, Thompson J et al (1993) Induction of human IgE synthesis in B cells by mast cells and basophils. *Nature* 365: 340–343

45 Pawankar R, Okuda M, Yssel H, Okumura K, Ra C (1997) Nasal mast cells in perennial allergic rhinitics exhibit increased expression of the Fc epsilonRI, CD40L, IL-4, and IL-13, and can induce IgE synthesis in B cells. *J Clin Invest* 99(7): 1492–1499

46 Nakajima T, Inagaki N, Tanaka H, Tanaka A, Yoshikawa M, Tamari M, Hasegawa K, Matsumoto K, Tachimoto H, Ebisawa M et al (2002) Marked increase in CC chemokine gene expression in both human and mouse mast cell transcriptomes following Fcepsilon receptor I cross-linking: an interspecies comparison. *Blood* 100(12): 3861–3868

47 Wakahara S, Fujii Y, Nakao T, Tsuritani K, Hara T, Saito H, Ra C (2001) Gene expression profiles for Fc epsilon RI, cytokines and chemokines upon Fc epsilon RI activation in human cultured mast cells derived from peripheral blood. *Cytokine* 16(4): 143–152

48 Rumsaeng V, Vliagoftis H, Oh CK, Metcalfe DD (1997) Lymphotactin gene expression in mast cells following Fc(epsilon) receptor I aggregation: modulation by TGF-beta, IL-4, dexamethasone, and cyclosporin A. *J Immunol* 158(3): 1353–1360

49 Baghestanian M, Hofbauer R, Kiener HP, Bankl HC, Wimazal F, Willheim M, Scheiner O, Fureder W, Muller MR, Bevec D et al (1997) The c-kit ligand stem cell factor and anti-IgE promote expression of monocyte chemoattractant protein-1 in human lung mast cells. *Blood* 90(11): 4438–4449

50 Packard KA, Khan MM (2003) Effects of histamine on Th1/Th2 cytokine balance. *Int Immunopharmacol* 3(7): 909–920

51 Stephens R, Randolph DA, Huang G, Holtzman MJ, Chaplin DD (2002) Antigen-nonspecific recruitment of Th2 cells to the lung as a mechanism for viral infection-induced allergic asthma. *J Immunol* 169(10): 5458–5467

52 Matsushima H, Yamada N, Matsue H, Shimada S (2004) TLR3-, TLR7-, and TLR9-mediated production of proinflammatory cytokines and chemokines from murine connective tissue type skin-derived mast cells but not from bone marrow-derived mast cells. *J Immunol* 173(1): 531–541

53 Kulka M, Alexopoulou L, Flavell RA, Metcalfe DD (2004) Activation of mast cells by

double-stranded RNA: evidence for activation through Toll-like receptor 3. *J Allergy Clin Immunol* 114(1): 174–182

54 Nagase H, Okugawa S, Ota Y, Yamaguchi M, Tomizawa H, Matsushima K, Ohta K, Yamamoto K, Hirai K (2003) Expression and function of Toll-like receptors in eosinophils: activation by Toll-like receptor 7 ligand. *J Immunol* 171(8): 3977–3982

55 Olszewska-Pazdrak B, Casola A, Saito T, Alam R, Crowe SE, Mei F, Ogra PL, Garofalo RP (1998) Cell-specific expression of RANTES, MCP-1, and MIP-1alpha by lower airway epithelial cells and eosinophils infected with respiratory syncytial virus. *J Virol* 72(6): 4756–4764

56 Ono SJ, Nakamura T, Miyazaki D, Ohbayashi M, Dawson M, Toda M (2003) Chemokines: roles in leukocyte development, trafficking, and effector function. *J Allergy Clin Immunol* 111(6): 1185–1199

57 Trifilo MJ, Lane TE (2004) The CC chemokine ligand 3 regulates $CD11c^{+}CD11b^{+}CD8alpha^{-}$ dendritic cell maturation and activation following viral infection of the central nervous system: implications for a role in T cell activation. *Virology* 327(1): 8–15

58 Trifilo MJ, Bergmann CC, Kuziel WA, Lane TE (2003) CC chemokine ligand 3 (CCL3) regulates CD8(+)-T-cell effector function and migration following viral infection. *J Virol* 77(7): 4004–4014

59 Loetscher P, Seitz M, Clark-Lewis I, Baggiolini M, Moser B (1996) Activation of NK cells by CC chemokines. Chemotaxis, $Ca2^{+}$ mobilization, and enzyme release. *J Immunol* 156(1): 322–327

60 Dorner BG, Scheffold A, Rolph MS, Huser MB, Kaufmann SH, Radbruch A, Flesch IE, Kroczek RA (2002) MIP-1alpha, MIP-1beta, RANTES, and ATAC/lymphotactin function together with IFN-gamma as type 1 cytokines. *Proc Natl Acad Sci USA* 99(9): 6181–6186

61 Tekkanat KK, Maassab H, Miller A, Berlin AA, Kunkel SL, Lukacs NW (2002) RANTES (CCL5) production during primary respiratory syncytial virus infection exacerbates airway disease. *Eur J Immunol* 32(11): 3276–3284

Chemokines as drug targets

Chemokines as drug targets

Amanda E.I. Proudfoot, Christine A. Power, Matthias Schwarz and Timothy N.C. Wells

Serono Pharmaceutical Research Institute, 14, Chemin des Aulx, 1228 Plan-les Ouates, Switzerland

Introduction

The question of how the migration of cells can be specifically controlled was first addressed at least 15 years ago with the identification of the neutrophil chemattractant, IL-8, and the monocyte attractants, MCP-1 and MIP-1α. The rapid expansion of this family of proteins, known as *chemo*attractant cyto*kines* and re-named chemokines, showed that the idea of one chemokine being responsible for the recruitment of a single cell type was over simplistic in most cases.

The identification of the first chemokine receptors however, indicated that this family would be 'druggable'. Most cytokines and growth factors generally act on receptors that are either single chain or heterodimeric but chemokines are the only cytokines so far known to act on seven transmembrane G protein-coupled receptors. The binding sites of the majority of cytokines and growth factors are relatively large, and have proved difficult to block with small molecule inhibitors [1]. In the case of chemokine receptors, previous work done on other G-protein coupled receptors has shown that it is possible to use a small molecule, acting inside the transmembrane regions, to stabilise the receptor in an inactive conformation, and thus produce small molecule antagonists.

The chemokine family thus opened a new series of receptors to target for anti-inflammatory therapies. It is well known that the 7TM receptor family is in fact the cornerstone of the pharmaceutical industry with a third of currently prescribed drugs targeted against members of this family. Therefore in the early 1990s, the hope was that inhibitors of the IL-8 receptors would provide therapies for acute inflammation and inhibitors of the first CC chemokine receptor, the shared RANTES/MIP-1α receptor, now known as CCR1, would be useful for the treatment of chronic inflammation. However, the plot thickened as the number of chemokine receptors identified grew rapidly. By 1996 there were 11 functional receptors and by

Chemokine Biology – Basic Research and Clinical Application, Volume II
edited by Kuldeep Neote, Gordon L. Letts and Bernhard Moser

the turn of the century the number had reached 18. Researchers in the chemokine field then had the task of identifying which receptors were involved in which disease processes – and more importantly to prove that the distinct family members were in fact, valid therapeutic targets.

The target validation process has involved several approaches, which are summarised briefly below. In order to identify the relevant target for a drug discovery program, it is essential to delineate which ligands/receptors are important in the disease of interest. Although a target is never validated until a drug based on the target has been successful in a human disease based clinical trial, it is still possible to build up a picture from a variety of sources.

Protein expression in the disease

Ideally this is measured directly at the protein level by immunohistochemistry or ELISA of human disease samples. Accessible biological fluids are often used such as serum, cerebrospinal fluid (CSF) and bronchiolar lavage (BAL) fluid. BAL isolated from the lungs of asthmatic patients has been shown to contain elevated levels of CCL5/RANTES [2] and CCL11/Eotaxin, CCL7/MCP-3 and CCL13/MCP-4 [3] implicating the receptors for these chemokines as potential targets. A more challenging task is obtaining tissue samples due to the limited availability of good quality human material. Two elegant studies were carried out on brain lesions isolated from autopsy samples taken from multiple sclerosis (MS) patients which indicated the presence of the inflammatory chemokines CCL5/RANTES, CCL2/MCP-1, CXCL10/IP-10 and CXCL9/MIG in the disease tissue [4, 5]. This implied that these chemokines and their receptors, CCR1 and CCR5, CCR2 and CXCR3, play a role in either the development or the resolution of the disease process. Many other such studies have been reported in the literature such as the high levels of CCL5/RANTES in psoriatic plaques, allowing its purification [6], elevated levels of CXCL8/IL-8 in pleurisy [7] and CCL11/Eotaxin in the serum of asthmatics [8].

mRNA levels that imply increased protein expression

Differential gene expression studies have been widely used in the identification of relevant targets. Despite their usefulness in highlighting the involvement of rare chemokines in disease, the most useful data to date has come from studies on purified cell populations [9]. The results from studies on biopsies or surgical samples which contain multiple cells types are often difficult to interpret unless very strict criteria have been applied to the samples used, e.g., age, sex and anatomical site matched tissues from patients with known treatment history and comparable disease status.

Neutralizing antibodies

Neutralization of MIP-1α and MCP-1 reduced inflammation in a rat model of experimental allergic encephalitis (EAE, a model for multiple sclerosis) [10]. This study demonstrated that MIP-1α was implicated in the onset of the disease, whilst MCP-1 played a key role in the relapses. These findings are corroborated by expression data for the ligand/receptor pair observed in human disease, as described above. Neutralization of CXCR3 prevented the rejection of cardiac transplants, showing the same phenotype as the $CXCR3^{-/-}$ mice – an excellent example of two approaches arriving at the same conclusion [11].

Genetic targeting of chemokine/chemokine receptors

Mice deficient in a particular ligand or receptor have proved to be an excellent approach to study the function of specific chemokine ligand-receptor interactions *in vivo*, particularly in models of inflammatory and infectious disease. Knockout of CCR1, CCR2 and JE (murine CCL2/MCP-1) support their potential as targets for therapeutic intervention in MS [12–15]. The major role of CCL2/MCP-1 and CCR2 in the pathogenesis of models of atherosclerosis was also confirmed using knockout mice [16, 17]. Possibly the most dramatic contribution to target validation of a chemokine receptor target has come from human genetic studies. Approximately 1% of the Caucasian population carries a homozygous deletion in the gene for CCR5, called Δ32 CCR5 [18]. This 32 base pair deletion in the CCR5 gene leads to production of a truncated protein which is not expressed on the cell surface – effectively producing a receptor knockout. Adult individuals, carrying this deletion, appear perfectly healthy, with the added advantage of being resistant to human immunodeficiency virus (HIV) infection. Thus, based on the human genetic data alone, it is likely that a selective CCR5 inhibitor would be sufficient to prevent infection by HIV, without having serious side effects.

It must be borne in mind that the prediction of validated targets is not always successful. Eosinophils are often the predominant inflammatory cell population present in allergic diseases and the unique functional receptor for CCL11/eotaxin, CCR3, is expressed at high levels on this cell type. While CCR3 and its ligand eotaxin were found to be highly expressed in the bronchial biopsies and BAL isolated from asthmatics [3], and serum eotaxin levels have been demonstrated to be a marker of disease severity in asthma patients [8], the mice in which these genes were deleted did not express the predicted phenotype, i.e., a resistance to experimental allergic asthma. The true contribution that eosinophils make to disease progression is, however, still constantly debated [19, 20] and therefore the initial optimism of finding small molecule inhibitors of CCR3, the chemokine receptor pivotal to asthma and allergy, has not produced a moiety entering clinical trials to date. Clinical studies with antibodies which block IL-5, the cytokine that mediates eosinophil differentiation and priming, have on the other hand shown that it is possible to reduce the circulating levels of eosinophils to normal levels, without having any effect on the disease outcome. This emphasizes the point that the presence of a cytokine or a cell type at the scene of the crime, can be pathological or protective depending on the disease.

Overall, data from several approaches used for target validation has led to the identification of certain receptors associated with specific diseases, which we have attempted to summarize in Figure 1, with the caveat that not all associations may be represented. In the remainder of this overview we will attempt to describe the 'State of the Art' of drugs aimed at the chemokine system both in clinical trials as well as in the research phase.

Pharmaceutical drug development: Small molecule inhibitors

The point of intervention that has been adopted by most companies is no doubt the inhibition of the ligand/receptor interaction, and the most sought after solution is orally available small molecule inhibitors. As mentioned above, the receptor class that is activated by chemokines is the 7TM, G protein-coupled receptor family – the most druggable target in the industry. However chemokines are not typical 7TM ligands, which are generally small molecules such as adrenalin, histamine, and acetylcholine – to name a few – and these ligands have molecular masses of approximately 200–400 Da – which are very small compared to the chemokine ligands of around 8,000 Da, which although small in protein terms is considered big for a 7 TM ligand. This identified a potential caveat comparable to the cytokine/receptor interaction – would a small molecule be able to inhibit the ligand/receptor interaction in a competitive inhibition mode? There was also a second problem for the majority of chemokine receptors – that of redundancy. Which ligand should be used in a high throughput screening campaign? One answer was actually obtained by the identification of inhibitory compounds, and well exemplified by one of the first small molecule inhibitors to be published in the literature – an inhibitor of CCR5, TAK779. This compound was able to inhibit all three ligands for CCR5 (CCL5/RANTES,

Figure 1
Chemokine receptor interactions and their association with disease
The familiar names for the chemokines are shown in the first column, and the systemic nomenclature is shown in the second. The association of receptors with the diseases shown on the right result from a cumulation of studies involving upregulation of RNA and protein in animal models of disease and human samples, as well as studies using mice in which the genes have been deleted.

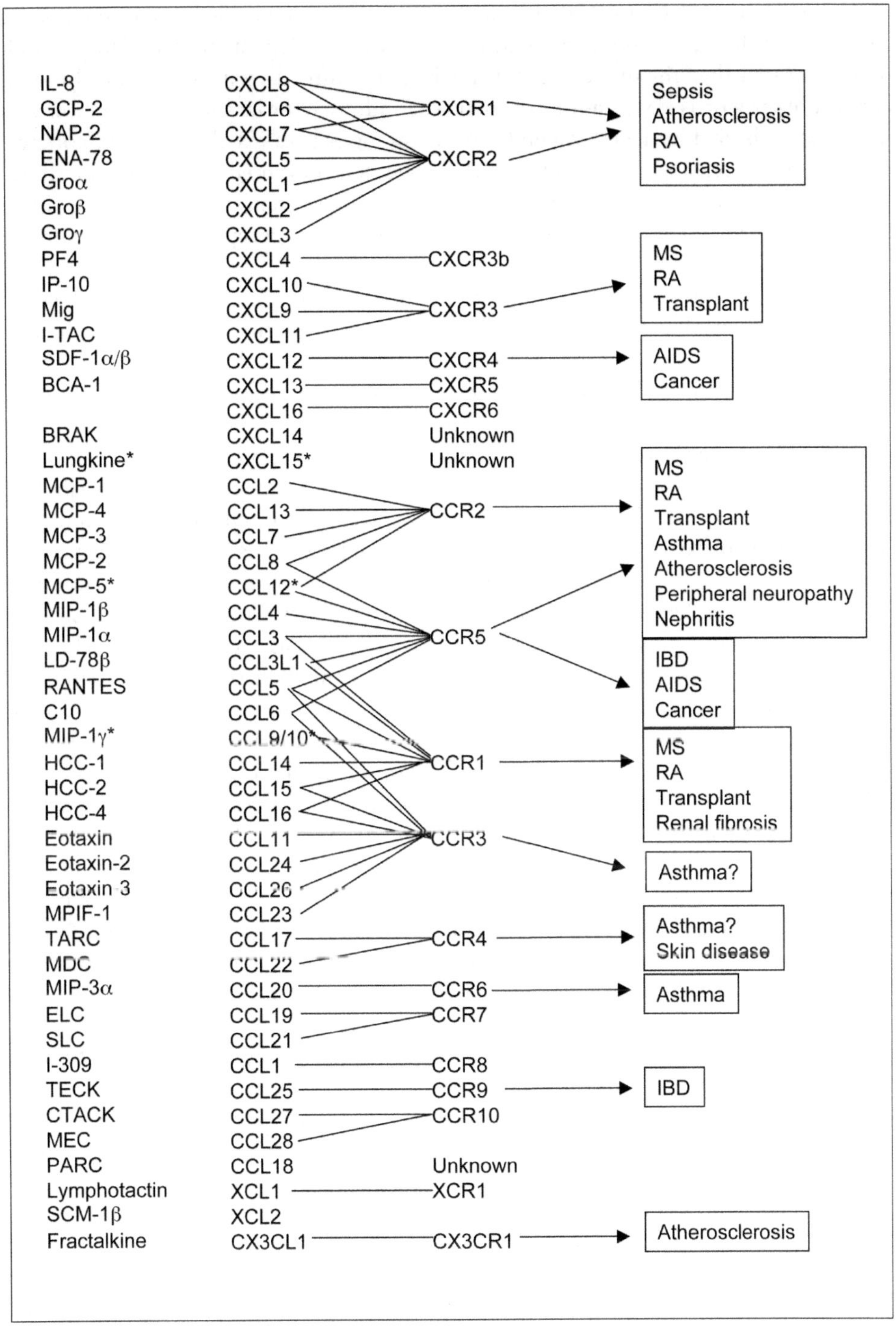
IL-8 CXCL8
GCP-2 CXCL6
NAP-2 CXCL7
ENA-78 CXCL5
Groα CXCL1
Groβ CXCL2
Groγ CXCL3
PF4 CXCL4
IP-10 CXCL10
Mig CXCL9
I-TAC CXCL11
SDF-1α/β CXCL12
BCA-1 CXCL13
CXCL16
BRAK CXCL14
Lungkine* CXCL15*
MCP-1 CCL2
MCP-4 CCL13
MCP-3 CCL7
MCP-2 CCL8
MCP-5* CCL12*
MIP-1β CCL4
MIP-1α CCL3
LD-78β CCL3L1
RANTES CCL5
C10 CCL6
MIP-1γ* CCL9/10*
HCC-1 CCL14
HCC-2 CCL15
HCC-4 CCL16
Eotaxin CCL11
Eotaxin-2 CCL24
Eotaxin 3 CCL26
MPIF-1 CCL23
TARC CCL17
MDC CCL22
MIP-3α CCL20
ELC CCL19
SLC CCL21
I-309 CCL1
TECK CCL25
CTACK CCL27
MEC CCL28
PARC CCL18
Lymphotactin XCL1
SCM-1β XCL2
Fractalkine CX3CL1
CXCR1
CXCR2
CXCR3b
CXCR3
CXCR4
CXCR5
CXCR6
Unknown
Unknown
CCR2
CCR5
CCR1
CCR3
CCR4
CCR6
CCR7
CCR8
CCR9
CCR10
Unknown
XCR1
CX3CR1
Sepsis
Atherosclerosis
RA
Psoriasis
MS
RA
Transplant
AIDS
Cancer
MS
RA
Transplant
Asthma
Atherosclerosis
Peripheral neuropathy
Nephritis
IBD
AIDS
Cancer
MS
RA
Transplant
Renal fibrosis
Asthma?
Asthma?
Skin disease
Asthma
IBD
Atherosclerosis

CCL3/MIP-1α and CCL4/MIP-1β) with nanomolar potency – but the interesting finding was that it was not a competitive inhibitor. An elegant study by Dragic et al. demonstrated that this molecule did not interfere directly with the ligand binding site (in other words it was not a competitive inhibitor) but its site of interaction was in a pocket buried in the transmembrane spanning helices – a classical non-competitive inhibitor causing a conformational change that prevented the binding of all ligands [21]. Subsequently, the finding that chemokine receptors do not have single binding sites for their ligands has been extended to other receptors, a good example being CXCR3 [22].

Watching the progress of the high throughput screening campaigns conducted by the pharmaceutical companies highlights the reduction in time between the identification of a target and patenting the first inhibitors – no doubt due to the huge advances in technologies used in these campaigns over the last decade. We have attempted to capture the success (and failures) in the identification of small molecule inhibitors that have been patented since the identification of the first chemokine receptors in 1991 [23, 24]. In the early 1990s, the main challenge consisted of designing small molecules able to function as antagonists of chemokine receptors *in vitro*, as illustrated by the scarce results obtained from many of the early high-throughput screening campaigns [25]. In fact, many in the field even doubted whether small molecule chemokine receptor antagonists would ever be found [1] but luckily this pessimism was unfounded and thus optimism was regained when the first inhibitors were reported in 1998 [26–28]. The number of published patents is summarized in Figure 2 and allows certain observations to be made. First, the identification of the chemokine receptors as co-receptors for HIV infectivity in 1996 served as a rapid accelerator in the identification of small molecule inhibitors by the pharmaceutical industry. Only 2 years later, patents for CCR5 inhibitors were published, whereas it took 6 years for the first inhibitors of CCR1 to appear. Second, certain receptors appear to be far more amenable for the identification of small molecule inhibitors. Again taking the HIV co-receptors, CCR5 and CXCR4 as an example – there are over 160 patents published for CCR5 to date, while for CXCR4 there are under 30. The number of patents for CCR2 inhibitors, one of the best validated receptors for inflammation is also under 30, whilst for CCR3, a favorite target for asthma, there are almost 70.

Once the first wave of molecules had been identified, chemists were able to decipher which structural determinants are required for small molecules to act as chemokine receptor antagonists [29], and have since developed numerous distinct chemical series of antagonists acting specifically on a wide variety of therapeutically relevant chemokine receptors [30]. A common trait quickly emerged in chemokine receptor inhibitors – they often encountered the same toxicity problems – interaction with the cardiac potassium channel hERG, and inhibition of certain cytochrome P450 isoforms, as well as lack of selectivity towards certain biogenic amine receptors, being the most frequently encountered problems. However, knowl-

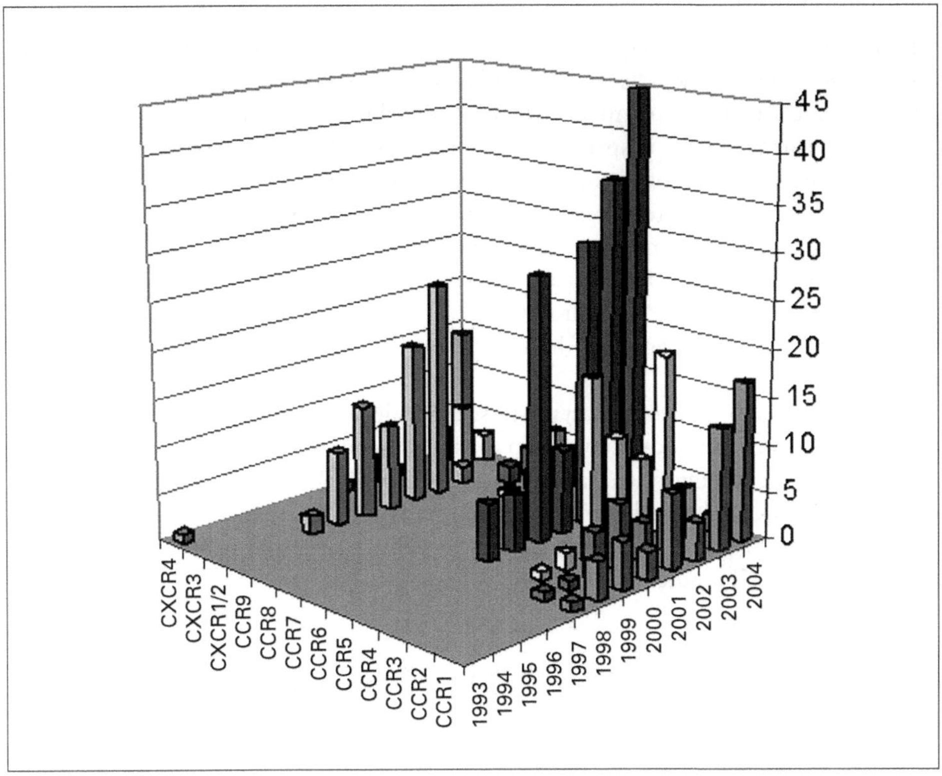

Figure 2
The number of patents published for small molecule inhibitors of chemokine receptors. The number of patents are shown published each year for individual chemokine receptors at the time of writing.

edge of these toxicity issues has in fact helped the medicinal chemistry programs as these are addressed very early on in development rather than later as was common practice for other programs.

Drug discovery of chemokine receptors faced another very important challenge – once the problem of potency *in vitro* and toxicity had been addressed, the next hurdle consisted of demonstrating efficacy *in vivo*, in appropriate animal models of disease. This was difficult for two reasons. The first is that of species selectivity – a problem which is encountered by many protein therapeutic programs. For small molecule inhibitors, the problem develops along with the progress made by the medicinal chemists. Initial hits, which have low affinities (probably in the high nanomolar range) will often inhibit both human and rodent receptors. However, as we are aiming at inhibiting the human receptors for treatment of human diseases, but are also constrained to prove the efficacy of our molecules in animal models – we are posed with the problem that as we increase the potency for the human receptor, we lose efficacy on the rodent receptor. Emphasis is placed on rodent species here, since animal models in non-human primates are exorbitantly expensive for compound screening. The problem of species specificity is not well documented in the literature, with the exception of the informative studies conducted on the CCR1 inhibitor, BX471 [31–34]. To overcome the species selectivity problem, certain researchers have chosen to perform the pharmacology studies with close analogues in the same chemical series showing residual activity on rodent receptors. The second major problem encountered, also common for small molecules, was the fact that some of the early classes of antagonists had poor physicochemical and pharmacokinetic properties.

However it is extremely gratifying to see that despite the many obstacles encountered by the pharmaceutical industry in the high-throughput screening programs, we are now seeing compounds targeting CCR1, CCR2, CCR3, CCR5, CXCR2, CXCR3 and CXCR4 that have successfully passed Phase I clinical trials in healthy volunteers and are now hopefully poised to deliver the final proof of therapeutic efficacy in Phase II studies in patients.

Biotechnology drug development: Protein therapeutics

The chemokine system offers potential targets to both the pharmaceutical and biotechnology industries. Although pharmaceutical companies have tended to concentrate on the search for small molecules, protein therapeutics also remains a very promising strategy. There are several strategies that have been adopted, including antibodies, modified chemokines, chimeric proteins and binding proteins.

Neutralizing monoclonal antibodies (mAbs), principally against the ligands, have been used extensively in animal models of disease, but only a few are being developed for therapeutic use – perhaps a reflection of the worry that mAbs would merely serve as a proof of concept for the emerging orally available small molecule receptor antagonists. This is rather regrettable since the development time for mAbs to reach the clinic can be much shorter than for small molecules, and moreover, mAbs generally have significantly fewer toxicity problems. However, they do suffer from the species selectivity problem since they can only be tested in non-human primates. An example is Millennium's humanized mAb targeting CCR2, which was tested in a model of restenosis in a non-human primate [35], prior to progression into man; it is currently being tested in Phase IIa studies for RA. The most advanced anti-chemokine antibody was an anti-CXCL8/IL-8 monoclonal, which was tested in a clinical trial of chronic obstructive pulmonary disease (COPD). It showed an improvement in the transition dyspnea index (TDI), but no improvement in lung function or health status of the patients was observed [36]. In a second trial, anti-

CXCL8/IL-8 failed to show efficacy in psoriasis. This underlines the risks in drug development. The finding of high levels of CXCL8/IL-8 in psoriatic skin led researchers to jump to the conclusion that blocking CXCL8/IL-8 activity would result in a reversal of the disease state. This has not been the outcome observed in clinical trials. To our knowledge, there are no mAbs directed to chemokine ligands that are currently being tested in clinical trials in patients, although mAbs against CCL11/Eotaxin and CCL2/MCP-1 have entered preclinical development.

There are extensive reports on the efficacy of modified chemokines in animal models of disease. These include truncated chemokines, such as (9-68) MCP-1 [37] and 7ND-MCP-1 [38], which have been tested both by administration of the recombinant protein or by gene therapy approaches [39–41]. Similarly, modifications of the amino terminus by extension such as Met-RANTES or by chemical modifications according to the prototype AOP-RANTES have also showed promising therapeutic potential. As far as we are aware, Met-RANTES is not being developed for the clinic, presumably for strategic business reasons, despite the ever increasing numbers of reports on its efficacy in various disease models (summarized in part in a previous review [42]). This is perhaps due to the fact that it retains partial agonist activity [43]. Improved variants of the prototype AOP-RANTES, initially shown to have extraordinary anti-HIV infectivity properties [44], have recently been shown to be efficacious in a non-human primate model in providing protection against infection [45, 46]. Although PSC RANTES was able to prevent infection of cells at nanomolar concentrations *in vitro*, *in vivo* the concentrations required to significantly reduce infection were in the millimolar range. The reasons for this large discrepancy remain to be resolved.

The essential interaction of chemokines with cell surface glycosaminoglycans (GAGs) [47–49] presents another strategy of interfering with chemokine activity. This interaction was shown to be essential for chemokine-induced cellular recruitment *in vivo* [50] and interference with the GAG interaction by the RANTES variant, [^{44}AANA^{47}]-RANTES was shown to have anti-inflammatory properties in cellular recruitment models which transcribed into anti-inflammatory properties in a disease model for multiple sclerosis, EAE [51]. Interestingly the apparent mechanism of action of this molecule is the disruption of the wild type RANTES oligomeric composition, which appears to be obligatory for its activity [52–54].

An interesting approach has been adopted by the group led by Matthias Mack. Fusion of the N-terminus of a truncated form of *Pseudomonas* exotoxin A to RANTES/CCL5 produced a molecule that allows internalization of the RANTES receptors through the endocytotic process [55]. The proof of concept of this approach has been demonstrated by efficacy in a murine model of asthma induced by fungal infection [56].

The last approach used to block chemokine/chemokine receptor interactions reported here is that of chemokine binding proteins. The human genome has evolved several endogenous strategies to combat inflammation in the guise of binding proteins that neutralize proinflammatory cytokines. Examples of these are IL-1Rα [57], TBP-I and II (TNF binding proteins) [58] and IL-18BP (interleukin-18 binding protein) [59]. Similarly viruses have evolved to produce such proteins and their genomes are known to encode binding proteins that neutralize cytokines such as IFN-γ and IL-18, as well as chemokines. Several virally encoded chemokine binding proteins have been described, and the reader is referred to a specialist review for a detailed description [60]. Two pox virus encoded chemokine binding proteins, MT-1, which inhibits receptor activation, and MT-7, believed to prevent chemokine binding to glycosaminoglycans (GAGs), have been tested along with the Herpes encoded binding protein, M3, in a rat aortic allograft transplantation model. Single administrations showed dose dependent activity at surprisingly low doses ranging from 5–5000 pg with remarkable efficacy in preventing intimal hyperplasia and vasculopathy, correlating with decreased mononuclear cell recruitment [61]. These findings are supported by the use of an inducible M3 transgenic mouse which demonstrated a significant reduction in intimal hyperplasia, critical in the development of atherosclerosis [62]. These experiments highlight the potential of these molecules for acute indications. The immunogenic effects of these agents are unknown and need to be well characterized before considering their use in the treatment of chronic diseases. More recently, an anti-CXCL8/IL-8 activity has been described in tick saliva [63]. This protein may play an important role in protecting the parasite from the host immune response, allowing them to feed on the host for relatively long periods (up to weeks) at a time. This mechanism offers a potential use for these proteins as anti-inflammatory molecules. A human chemokine binding protein has recently appeared in the patent literature (WO2004055050). This protein is distinct from the viral chemokine binding proteins in that it is a functional domain of a known nuclear protein. Efficacy and specificity in a collagen induced arthritis model are reported for this molecule in the patent.

Perspectives

On a final note, it is over 15 years since the molecular characterization of the first chemokines, IL-8 and MCP-1. In that period we have seen the family blossom to include almost 50 cytokines, and 20 receptors. The use of genetically modified mouse models, combined with detection of chemokine/receptor expression in human disease, has now given us a much clearer understanding of the numerous roles of the various chemokines. From a therapeutic viewpoint however, it is fair to say that progress has been slow. Despite the fact that chemokine receptors are G-protein coupled receptors, they have generally proven to be remarkably refractory to small molecule approaches. However, a number of alternative therapeutic strategies targeting the chemokine system have now reached the stage where they are being tested in a clinical setting, in a range of human diseases. The fruition of some of these approaches into new medicines – opening up another front in our fight against disease – is eagerly anticipated.

References

1 McKinnon M, Proudfoot AEI, Wells TNC, Solari R (1996) Strategies for the discovery of cytokine receptor antagonists. *Drug News Perspect* 9: 389–398

2 Teran LM, Noso N, Carroll M, Davies DE, Holgate S, Schroder JM (1996) Eosinophil recruitment following allergen challenge is associated with the release of the chemokine RANTES into asthmatic airways. *J Immunol* 157: 1806–1812

3 Ying S, Meng Q, Zeibecoglou K, Robinson DS, Macfarlane A, Humbert M, Kay AB (1999) Eosinophil chemotactic chemokines (eotaxin, eotaxin-2, RANTES, monocyte chemoattractant protein-3 (MCP-3), and MCP-4), and C-C chemokine receptor 3 expression in bronchial biopsies from atopic and nonatopic (Intrinsic) asthmatics. *J Immunol* 163: 6321–6329

4 Sorensen TL, Tani M, Jensen J, Pierce V, Lucchinetti C, Folcik VAQ, Rottman J, Sellebjerg F, Strieter RM, Frederiksen JLR (1999) Expression of specific chemokines and chemokine receptors in the central nervous system of multiple sclerosis patients. *J Clin Invest* 103: 807–815

5 Balashov KE, Rottman JB, Weiner HL, Hancock WW (1999) CCR5(+) and CXCR3(+) T cells are increased in multiple sclerosis and their ligands MIP-1 alpha and IP-10 are expressed in demyelinating brain lesions. *Proc Natl Acad Sci USA* 96: 6873–6878

6 Schroder JM (1997) Identification and structural characterization of chemokines in lesional skin material of patients with inflammatory skin disease. *Methods Enzymol* 288: 266–297

7 Broaddus VC, Hebert CA, Vitangcol RV, Hoeffel JM, Bernstein MS, Boylan AM (1992) Interleukin-8 is a major neutrophil chemotactic factor in pleural liquid of patients with empyema. *Am Rev Respir Dis* 146: 825–830

8 Nakamura H, Weiss ST, Israel E, Luster AD, Drazen JM, Lilly CM (1999) Eotaxin and impaired lung function in asthma. *Am J Respir Crit Care Med* 160: 1952–1956

9 Renzoni EA, Abraham DJ, Howat S, Shi-Wen X, Sestini P, Bou-Gharios G, Wells AU, Veeraraghavan S, Nicholson AG, Denton CP et al (2004) Gene expression profiling reveals novel TGFbeta targets in adult lung fibroblasts. *Respir Res* 5: 24

10 Kennedy KJ, Strieter RM, Kunkel SL, Lukacs NW, Karpus WJ (1998) Acute and relapsing experimental autoimmune encephalomyelitis are regulated by differential expression of the CC chemokines macrophage inflammatory protein-1 alpha and monocyte chemotactic protein-1. *J Neuroimmunol* 92: 98–108

11 Hancock WW, Lu B, Gao W, Csizmadia V, Faia K, King JA, Smiley ST, Ling M, Gerard NP, Gerard C (2000) Requirement of the chemokine receptor CXCR3 for acute allograft rejection. *J Exp Med* 192: 1515–1520

12 Rottman JB, Slavin AJ, Silva R, Weiner HL, Gerard CG, Hancock WW (2000) Leuko-

cyte recruitment during onset of experimental allergic encephalomyelitis is CCR1 dependent. *Eur J Immunol* 30: 2372–2377

13 Fife BT, Huffnagle GB, Kuziel WA, Karpus WJ (2000) CC chemokine receptor 2 is Critical for induction of experimental autoimmune. *J Exp Med* 192: 899–905

14 Izikson L, Klein RS, Charo IF, Weiner HL, Luster AD (2000) Resistance to experimental autoimmune encephalomyelitis in mice lacking the CC chemokine receptor (CCR)2. *J Exp Med* 192: 1075–1080

15 Huang D, Wang J, Kivisakk P, Rollins BJ, Ransohoff RM (2001) Absence of monocyte chemoattractant protein 1 in mice leads to decreased local macrophage recruitment and antigen-specific t helper cell type 1 immune response in experimental autoimmune encephalomyelitis. *J Exp Med* 193: 713–726

16 Peters W, Charo IF (2001) Involvement of chemokine receptor 2 and its ligand, monocyte chemoattractant protein-1, in the development of atherosclerosis: lessons from knockout mice. *Curr Opin Lipidol* 12: 175–180

17 Gosling J, Slaymaker S, Gu L, Tseng S, Zlot CH, Young SG, Rollins BJ, Charo IF (1999) MCP-1 deficiency reduces susceptibility to atherosclerosis in mice that overexpress human apolipoprotein B. *J Clin Invest* 103: 773–778

18 Samson M, Libert F, Doranz BJ, Rucker J, Liesnard C, Farber CM, Saragosti S, Lapoumeroulie C, Cognaux J, Forceille C et al (1996) Resistance to HIV-1 infection in caucasian individuals bearing mutant alleles of the CCR-5 chemokine receptor gene [see comments]. *Nature* 382: 722–725

19 Humbles AA, Lloyd CM, McMillan SJ, Friend DS, Xanthou G, McKenna EE, Ghiran S, Gerard NP, Yu C, Orkin SH et al (2004) A critical role for eosinophils in allergic airways remodeling. *Science* 305: 1776–1779

20 Lee JJ, Dimina D, Macias MP, Ochkur SI, McGarry MP, O'Neill KR, Protheroe C, Pero R, Nguyen T, Cormier SA et al (2004) Defining a link with asthma in mice congenitally deficient in eosinophils. *Science* 305: 1773–1776

21 Dragic T, Trkola A, Thompson DA, Cormier EG, Kajumo FA, Maxwell E, Lin SW, Ying W, Smith SO, Sakmar TP et al (2000) A binding pocket for a small molecule inhibitor of HIV-1 entry within the transmembrane helices of CCR5. *Proc Natl Acad Sci USA* 97: 5639–5644

22 Cox MA, Jenh CH, Gonsiorek W, Fine J, Narula SK, Zavodny PJ, Hipkin RW (2001) Human interferon-inducible 10-kDa protein and human interferon-inducible T cell alpha chemoattractant are allotopic ligands for human CXCR3: differential binding to receptor states. *Mol Pharmacol* 59: 707–715

23 Holmes WE, Lee J, Kuang WJ, Rice GC, Wood WI (1991) Structure and functional expression of a human interleukin-8 receptor. *Science* 253: 1278–1280

24 Murphy PM, Tiffany HL (1991) Cloning of complementary DNA encoding a functional human interleukin-8 receptor. *Science* 253: 1280–1283

25 Mellor GW, Fogarty SJ, OBrien MS, Congreve M, Banks MN, Mills KM, Jefferies B, Houston JG (1997) Searching for chemokine receptor binding antagonists by high throughput screening. *J Biomolecular Screening* 2: 153–157

26 White JR, Lee JM, Young PR, Hertzberg RP, Jurewicz AJ, Chaikin MA, Widdowson K, Foley JJ, Martin LD, Griswold DE et al (1998) Identification of a potent, selective non-peptide CXCR2 antagonist that inhibits interleukin-8-induced neutrophil migration. *J Biol Chem* 273: 10095–10098

27 Solari R, Offord RE, Remy S, Aubry JP, Wells TN, Whitehorn E, Oung T, Proudfoot AE (1997) Receptor-mediated endocytosis of CC-chemokines. *J Biol Chem* 272: 9617–9620

28 Hesselgesser J, Ng HP, Liang M, Zheng W, May K, Bauman JG, Monahan S, Islam I, Wei GP, Ghannam A et al (1998) Identification and characterization of small molecule functional antagonists of the CCR1 chemokine receptor. *J Biol Chem* 273: 15687–15692

29 Schwarz MK, Wells TN (2002) New therapeutics that modulate chemokine networks. *Nat Rev Drug Discov* 1: 347–358

30 Gao Z, Metz WA (2003) Unraveling the chemistry of chemokine receptor ligands. *Chem Rev* 103: 3733–3752

31 Liang M, Mallari C, Rosser M, Ng HP, May K, Monahan S, Bauman JG, Islam I, Ghannam A, Buckman B et al (2000) Identification and characterization of a potent, selective, and orally active antagonist of the CC chemokine receptor-1. *J Biol Chem* 275: 19000–19008

32 Liang M, Rosser M, Ng HP, May K, Bauman JG, Islam I, Ghannam A, Kretschmer PJ, Pu H, Dunning L et al (2000) Species selectivity of a small molecule antagonist for the CCR1 chemokine receptor. *Eur J Pharmacol* 389: 41–49

33 Horuk R, Shurey S, Ng HP, May K, Bauman JG, Islam I, Ghannam A, Buckman B, Wei GP, Xu W et al (2001) CCR1-specific non-peptide antagonist: efficacy in a rabbit allograft rejection model. *Immunol Lett* 76: 193–201

34 Anders HJ, Vielhauer V, Frink M, Linde Y, Cohen CD, Blattner SM, Kretzler M, Strutz F, Mack M, Grone HJ et al (2002) A chemokine receptor CCR-1 antagonist reduces renal fibrosis after unilateral ureter ligation. *J Clin Invest* 109: 251–259

35 Horvath C, Welt FG, Nedelman M, Rao P, Rogers C (2002) Targeting CCR2 or CD18 inhibits experimental in-stent restenosis in primates: inhibitory potential depends on type of injury and leukocytes targeted. *Circ Res* 90: 488–494

36 Mahler DA, Huang S, Tabrizi M, Bell GM (2004) Efficacy and safety of a monoclonal antibody recognizing interleukin-8 in COPD: a pilot study. *Chest* 126: 926–934

37 Gong JH, Clark-Lewis I (1995) Antagonists of monocyte chemoattractant protein 1 identified by modification of functionally critical NH2-terminal residues. *J Exp Med* 181: 631–640

38 Zhang Y, Rollins BJ (1995) A dominant negative inhibitor indicates that monocyte chemoattractant protein 1 functions as a dimer. *Mol Cell Biol* 15: 4851–4855

39 Gong JH, Ratkay LG, Waterfield JD, Clark LI (1997) An antagonist of monocyte chemoattractant protein 1 (MCP-1) inhibits arthritis in the MRL-lpr mouse model. *J Exp Med* 186: 131–137

40 Wada T, Furuichi K, Sakai N, Iwata Y, Kitagawa K, Ishida Y, Kondo T, Hashimoto H,

Ishiwata Y, Mukaida N et al (2004) Gene therapy via blockade of monocyte chemoattractant protein-1 for renal fibrosis. *J Am Soc Nephrol* 15: 940–948

41 Kitamoto S, Egashira K (2002) Gene therapy targeting monocyte chemoattractant protein-1 for vascular disease. *J Atheroscler Thromb* 9: 261–265

42 Proudfoot AE (2002) Chemokine receptors: multifaceted therapeutic targets. *Nat Rev Immunol* 2: 106–115

43 Proudfoot AE, Buser R, Borlat F, Alouani S, Soler D, Offord RE, Schroder JM, Power CA, Wells TN (1999) Amino-terminally modified RANTES analogues demonstrate differential effects on RANTES receptors. *J Biol Chem* 274: 32478–32485

44 Simmons G, Clapham PR, Picard L, Offord RE, Rosenkilde MM, Schwartz TW, Buser R, Wells TNC, Proudfoot AEI (1997) Potent inhibition of HIV-1 infectivity in macrophages and lymphocytes by a novel CCR5 antagonist. *Science* 276: 276–279

45 Kawamura T, Bruse SE, Abraha A, Sugaya M, Hartley O, Offord RE, Arts EJ, Zimmerman PA, Blauvelt A (2004) PSC-RANTES blocks R5 human immunodeficiency virus infection of Langerhans cells isolated from individuals with a variety of CCR5 diplotypes. *J Virol* 78: 7602–7609

46 Lederman MM, Veazey RS, Offord R, Mosier DE, Dufour J, Mefford M, Piatak M Jr., Lifson JD, Salkowitz JR, Rodriguez B et al (2004) Prevention of vaginal SHIV transmission in rhesus macaques through inhibition of CCR5. *Science* 306: 485–487

47 Rot A (1993) Neutrophil attractant/activation protein-1 (interleukin-8) induces *in vitro* neutrophil migration by haptotactic mechanism. *Eur J Immunol* 23: 303–306

48 Hoogewerf AJ, Kuschert GS, Proudfoot AE, Borlat F, Clark-Lewis I, Power CA, Wells TN (1997) Glycosaminoglycans mediate cell surface oligomerization of chemokines. *Biochemistry (Mosc)* 36: 13570–13578

49 Kuschert GS, Coulin F, Power CA, Proudfoot AE, Hubbard RE, Hoogewerf AJ, Wells TN (1999) Glycosaminoglycans interact selectively with chemokines and modulate receptor binding and cellular responses. *Biochemistry (Mosc)* 38: 12959–12968

50 Proudfoot AEI, Handel TM, Johnson Z, Lau EK, LiWang P, Clark L, Borlat F, Wells TNC, Kosco-Vilbois MH (2003) Glycosaminoglycan binding and oligomerization are essential for the *in vivo* activity of certain chemokines. *Proc Natl Acad Sci USA* 100: 1885–1890

51 Johnson Z, Kosco-Vilbois MH, Herren S, Cirillo R, Muzio V, Zaratin P, Carbonatto M, Mack M, Smailbegovic A, Rose M et al (2004) Interference with heparin binding and oligomerization creates a novel anti-inflammatory strategy targeting the chemokine system. *J Immunol* 173: 5776–5785

52 Baltus T, Weber KS, Johnson Z, Proudfoot AE, Weber C (2003) Oligomerization of RANTES is required for CCR1-mediated arrest but not CCR5-mediated transmigration of leukocytes on inflamed endothelium. *Blood* 102: 1985–1988

53 Shaw JP, Johnson Z, Borlat F, Zwahlen C, Kungl AJ, Roulin K, Harrenga A, Wells TN, Proudfoot AE (2004) The X-ray structure of RANTES: heparin-derived disaccharides allows the rational design of chemokine inhibitors. *Structure* 12: 2081–2093

54 Wagner L, Yang OO, Garcia-Zepeda EA, Ge Y, Kalams SA, Walker BD, Pasternack MS,

Luster AD (1998) β-Chemokines are released from HIV-1-specific cytolytic granules complexed to potreoglycans. *Nature* 391: 908–911
55 Bruhl H, Cihak J, Stangassinger M, Schlondorff D, Mack M (2001) Depletion of CCR5-expressing cells with bispecific antibodies and chemokine toxins: a new strategy in the treatment of chronic inflammatory diseases and HIV. *J Immunol* 166: 2420–2426
56 Schuh JM, Blease K, Bruhl H, Mack M, Hogaboam CM (2003) Intrapulmonary targeting of RANTES/CCL5-responsive cells prevents chronic fungal asthma. *Eur J Immunol* 33: 3080–3090
57 Eisenberg SP, Brewer MT, Verderber E, Heimdal P, Brandhuber BJ, Thompson RC (1991) Interleukin 1 receptor antagonist is a member of the interleukin 1 gene family: evolution of a cytokine control mechanism. *Proc Natl Acad Sci USA* 88: 5232–5236
58 Seckinger P, Dayer JM (1992) Natural inhibitors of TNF. *Immunol Ser* 56: 217–236
59 Novick D, Kim SH, Fantuzzi G, Reznikov LL, Dinarello CA, Rubinstein M (1999) Interleukin-18 binding protein: a novel modulator of the Th1 cytokine response. *Immunity* 10: 127–136
60 Alcami A (2003) Viral mimicry of cytokines, chemokines and their receptors. *Nat Rev Immunol* 3: 36–50
61 Liu L, Dai E, Miller L, Seet B, Lalani A, Macauley C, Li X, Virgin HW, Bunce C, Turner P et al (2004) Viral chemokine-binding proteins inhibit inflammatory responses and aortic allograft transplant vasculopathy in rat models. *Transplantation* 77: 1652–1660
62 Pyo R, Jensen KK, Wiekowski MT, Manfra D, Alcami A, Taubman MB, Lira SA (2004) Inhibition of intimal hyperplasia in transgenic mice conditionally expressing the chemokine-binding protein M3. *Am J Pathol* 164: 2289–2297
63 Hajnicka V, Kocakova P, Slavikova M, Slovak M, Gasperik J, Fuchsberger N, Nuttall PA (2001) Anti-interleukin-8 activity of tick salivary gland extracts. *Parasite Immunol* 23: 483–489

Screening and characterization of cyclic pentapeptide CXCR4 antagonists/inverse agonists using a pheromone responsive reporter gene in *Saccharomyces cerevisiae*: Utility of G protein coupled receptor constitutively active mutants

Zi-xuan Wang[1], Hirokazu Tamamura[2], Nicole Frilot[1], James Broach[3], Nobutaka Fujii[2], Stephen C. Peiper[1]

[1]Immunotherapy Center and Department of Pathology, Medical College of Georgia, Augusta, GA, USA; [2]Kyoto University, Kyoto, Japan; [3]Princeton University, Princeton, NJ, USA

Introduction

CXCR4 is a chemokine receptor that belongs to the GPCR superfamily. It specifically transduces the signal of one CXC chemokine ligand, stromal cell derived factor 1 [1, 2], which has been designated CXCL12 in the current nomenclature assigned to members of the chemoattractant cytokine gene family [3]. This GPCR-ligand pair plays a critical role in programming the directed migration of multiple cell types during embryologic development, including neural, endothelial, and hematopoietic progenitors, as well as primordial germ cells [4]. Mice rendered nullizygous for either CXCR4 or CXCL12 by gene targeting die *in utero* with incomplete formation of the cerebellum, vascular system and the medullary hematopoietic compartment [5–7]. During adult life, CXCL12 is a chemoattractant for multiple subsets of leukocyte and exposure to CXCR4 antagonists results in mobilization of (hematopoietic) stem cells [8]. Rare individuals carrying a mutation that results in loss of the normal cytoplasmic tail, either by truncation or translational frame shift, which precludes desensitization following activation have a dominant syndrome characterized by warts, hypogammaglobulinemia, infections, and myelokathexis (retention of myeloid cells in the bone marrow) [9].

In addition to its critical roles in embryogenesis and adult life, CXCR4 has multiple functions in pathologic physiology. It was the first coreceptor described for the CD4-dependent envelope-mediated entry of target cells by HIV-1 [10], and is required for infection by T-tropic strains spawned during late phases of infection. In addition, it also can be utilized by dual tropic species [11] that use both

Chemokine Biology – Basic Research and Clinical Application, Volume II
edited by Kuldeep Neote, Gordon L. Letts and Bernhard Moser

CCR5, the exclusive coreceptor for commonly transmitted forms, and CXCR4, which may evolve during this shift in tropism [12]. CXCL12 and CXCR4 antagonists block its utilization as an HIV-1 coreceptor [1, 2] and two inhibitory compounds, AMD3100 and ALX40-4C, have been tested in clinical trials in late stage AIDS patients, unfortunately with a minimal impact on disease course [13, 14]. The administration of AMD3100 was associated with some cardiotoxicity [14]. CXCR4 has also been found to play a role in programming the directed migration of tumor cells [15, 16] to target organs that secrete CXCL12. Analysis of human tumor tissues and cell lines reveals that many, including carcinomas of the breast [17], prostate [18], lung [19], urinary bladder [20], thyroid [21] and head and neck (squamous cell) [22], glioblastomas [23], neuroblastomas [24], rhabdomyosarcomas [25], chronic lymphocytic leukemias [26] and acute myelogenous leukemias [27] express CXCR4 that activates cytosolic signaling pathways and chemotaxis in response to exposure to CXCL12. Monoclonal antibodies to CXCR4 [15] and CXCR4 antagonists, including AMD3100 [17] and T140 [28], have been shown to block the metastasis of human tumor cell lines in xenograft models in immune deficient mice. Expression microarray analysis of highly metastatic variants of human breast cancer cell lines derived by biologic selection revealed that CXCR4 plays a key role in the metastatic phenotype [16]. Administration of CXCR4 antagonists has been found to induce mobilization of hematopoietic stem cells [8].

Since blockade of CXCR4 appears to be a potential therapy of late stage HIV-1 infection and to block metastatic spread of a variety of malignancies, multiple groups have focused on the development of candidate compounds that inhibit this molecular target. Currently, three approaches have been published to develop an effective CXCR4 antagonist that could be used in clinical trials. New generation compounds are being designed from the AMD3100 lead compound, which is not orally bioavailable and was associated with cardiotoxicity [14]. Another lead compound is KRH1636, which can be absorbed from the duodenum, but is not orally bioavailable [29].

Here we describe a new screening strategy to characterize the downsizing and optimization of cyclic pentapeptides designed from T140 [30], a polypeptide derivative of a naturally occurring horseshoe crab protein that is a CXCR4 antagonist. Human CXCR4 was expressed in *Saccharomyces cerevisiae* coupled to the pheromone response pathway to develop an efficient screening system for CXCR4 antagonists. The utilization of constitutively active variants of CXCR4 (CXCR4-CAM), which autonomously and continuously trigger pheromone responsive *HIS3* and *lacZ* reporter genes enabled us to categorize the pharmacologic properties of the available CXCR4 antagonists, and demonstrate that AMD3100 and ALX40-4C are weak partial agonists and T140 and derivatives are inverse agonists [31]. This system was used to screen a combinatorial library of cyclic pentapeptides composed of four amino acid residues critical to the T140 inverse agonist pharmacophore with a glycine spacer in spatial and chiral permutations. This approach enabled us to develop a lead compound and two subsequent generations. The biochemical and biological properties of these compounds will be discussed. Ultimately, the best predictor of biologic potency was the affinity of the compound for a specific CXCR4-CAM. The use of constitutively active GPCR variants may allow for screening of compounds for reactivity with stable active states of the receptor.

Materials and methods

Plasmids and yeast strains

The open reading frames encoding the wild type human CXCR4 and a constitutively active mutant were cloned in the Cp4258 vector and expressed along with a plasmid encoding the *FUS1-lacZ* reporter gene in the *CY12946* yeast strain as previously described [31]. Constructs were introduced into yeast cells using the Frozen-EZ Yeast Transformation-II kit (Zymo Research, Orange, CA).

Screening of cyclic pentapeptide libraries

Compounds were dissolved in 50% DMSO to a stock concentration of 5 mM and tested at a final concentration of 10 μM or other specified concentrations. Yeast strains expressing native CXCR4 or the CXCR4-CAM were expanded as previously described [31]. Aliquots of fresh cultures at OD600 ≈ 0.2 were incubated in the presence of candidates in 96 well plates for 6 h. For screens using the native receptor, CXCL12 was added to the culture at a final concentration of 2 μM along with the candidate compounds and incubated at 30°C for 6 h. Aliquots were solubilized and tested for expression of the β-galactosidase product of the pheromone-responsive *lacZ* reporter gene, using fluorescein di-β-D-galactopyranoside (FDG), a fluorescent β-galactosidase substrate (Molecular Probes, Eugene, OR). Enzymatic activity was determined using standard approaches. The experimental data were normalized using β-galactosidase activity values of native CXCR4 activated by CXCL12 or the basal activity of the CXCR4-CAM.

[^{125}I]CXCL12 binding

CHO transfectants stably expressing CXCR4 variants were incubated with 0.1 nM [^{125}I]CXCL12 (PerkinElmer Life Sciences) in the presence or absence of cold inhibitors using standard techniques as described previously [31].

Cytosolic calcium mobilization

For calcium flux experiments, CHO-CXCR4 transfectants expressing native human CXCR4 were loaded with Fura-2 acetoxymethyl ester (2 μg/ml) (Molecular Probes) as described previously. The response to CXCL12 was recorded using a spectrofluorometer (F2500, Hitachi, San Jose, CA) as described previously [31].

Chemotaxis assay

CHO transfectants were starved in MEMα medium containing 0.5% bovine serum albumin for 2 h. The cells were then washed and resuspended in the same medium at a concentration of 2×10^6 cells/ml. 100 μl of this cell suspension was added to the top chamber of 24-well transwell apparatus (6.5 mm diameter, 8.0 μm pore size; Corning Inc, Corning, NY). Agonist and antagonist were added to the lower chamber and the plates were incubated for 4 h at 37°C. Cells were fixed and stained with crystal violet in 20% ethanol. Cells retained in the top well of the insert were wiped off with a folded tissue paper. Cells that migrated to the bottom side of the insert membrane photographed using an inverted microscope and four random fields were manually counted. Chemotaxis was determined as a ratio of cells migrating in response to CXCL12 alone. The results are representative of three independent experiments.

Results

Screening of combinatorial cyclic pentapeptides in yeast

A combinatorial synthetic approach was used to generate an array of cyclic peptapeptides composed of four critical residues from the T140 template (Arg, Arg, Tyr, and Napthyl-alanine) with a glycine spacer permuted spatially and chirally. The array of compounds was screened using yeast strains programmed to express native or constitutively active versions of human CXCR4 coupled to a pheromone-responsive *FUS1-lacZ* reporter gene. While exposure of native CXCR4 to recombinant human CXCL12 resulting in activation of the pheromone response pathway and can detect the presence of a neutral antagonist, a weak partial agonist, and an inverse agonist, incubation of candidates with the CXCR4-CAM can only identify the latter two pharmacologic types of inhibitors.

Exposure of yeast cells expressing CXCR4 to CXCL12 resulted in a significant increase in expression of the *FUS1-lacZ* reporter gene, and this increase in β-galactosidase activity was inhibited in the presence of T140, the prototypical inverse agonist (Fig. 1A). Screening of the array of cyclic pentapeptide candidates revealed that the presence of only one, FC068, resulted in a significant decrease in β-galactosidase activity following CXCL12 activation of CXCR4. A parallel analysis of the cyclic pentapeptide array using yeast cells expressing the CXCR4-CAM revealed a high basal level of β-galactosidase activity, inhibition of this activity by the T140 template compound, and a further increase in response to CXCL12 (Fig. 1B). Two cyclic pentapeptide candidates, FC001 and FC002, increased the constitutive β-galactosidase activity, combined with further dose response experiments (data not shown) suggesting that they were weak partial agonists. FC068, the same compound that blocked CXCL12 activation of the pheromone-responsive *lacZ* reporter gene in yeast cells expressing native CXCR4, decreased the basal β-galactosidase activity resulting from the CXCR4-CAM to baseline levels, indicating that it is an inverse agonist.

In summary, the strategy of screening candidates in parallel in yeast strains expressing either native CXCR4 activated by CXCL12 or the CXCR4-CAM resulted in the detection of three candidates, two weak partial agonists, one inverse agonist, and no neutral antagonists.

The inverse agonist candidate, FC068, was used as a template to develop new generations of compounds generated by permutation of the chirality of the four residues critical to T140 activity. This resulted in the identification of a second generation compound, FC092, and a third generation compound, FC131. A fourth generation compound was created by the addition of a protective methyl group on an Arg residue to produce FC122. Analysis of the dose dependence of the activity of FC068, FC092, FC131, and FC122 revealed that T140 is only slightly more potent than the third and fourth generation cyclic pentapeptides and confirmed that all four generation compounds shift the conformation of the CXCR4-CAM to the inactive state, thereby decreasing the expression of the pheromone-responsive *lacZ* reporter gene (Fig. 1C). There was an increase in the potency of the inverse agonist activity of the sequential generations of cyclic pentapeptides evident from incremental dose dependent decreases in the β-galactosidase activity of yeast strains expressing the CXCR4-CAM following exposure to the first through to the fourth generation compounds.

Characterization of the binding affinity of cyclic pentapeptide CXCR4 antagonists

The binding affinity of the four generations of compounds was determined by displacement of [^{125}I]CXCL12 binding to CHO transfectants stably expressing CXCR4. As shown in Figure 2, the hierarchy of binding affinities was T140 ≥ FC122 ≥ FC131 > FC092 > FC068, indicating an incremental decrease in the IC_{50} of each generation of compounds (IC_{50}: T140, 3.4 nM; FC068, 16.9 nM; FC092, 7.2 nM; FC131, 4.0 nM; FC122, 3.5 nM).

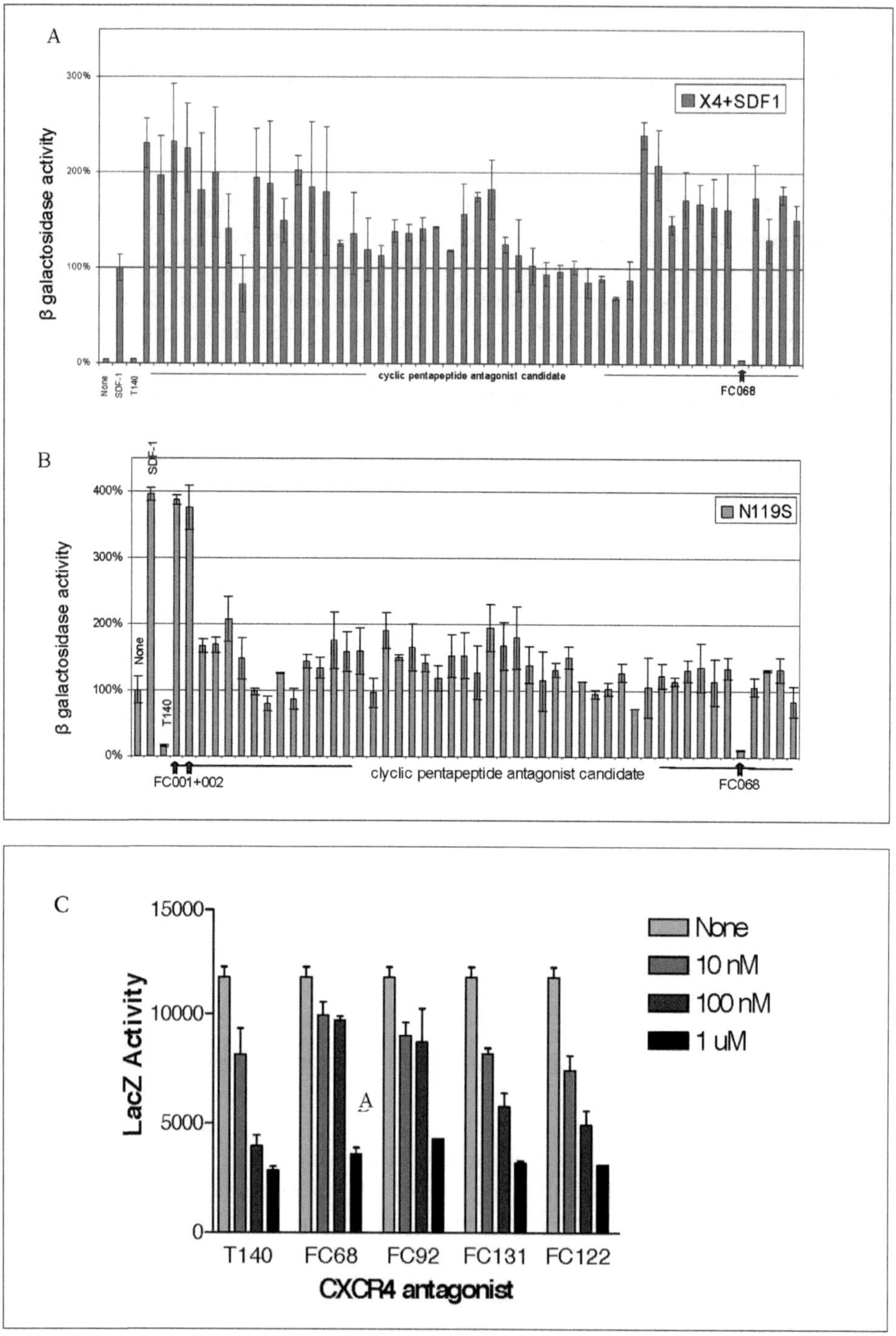
A
β galactosidase activity
300%
200%
100%
0%
X4+SDF1
None
SDF-1
T140
cyclic pentapeptide antagonist candidate
FC068
B
β galactosidase activity
400%
300%
200%
100%
0%
N119S
None
SDF-1
T140
FC001+002
clyclic pentapeptide antagonist candidate
FC068
C
LacZ Activity
15000
10000
5000
0
None
10 nM
100 nM
1 uM
T140
FC68
FC92
FC131
FC122
CXCR4 antagonist

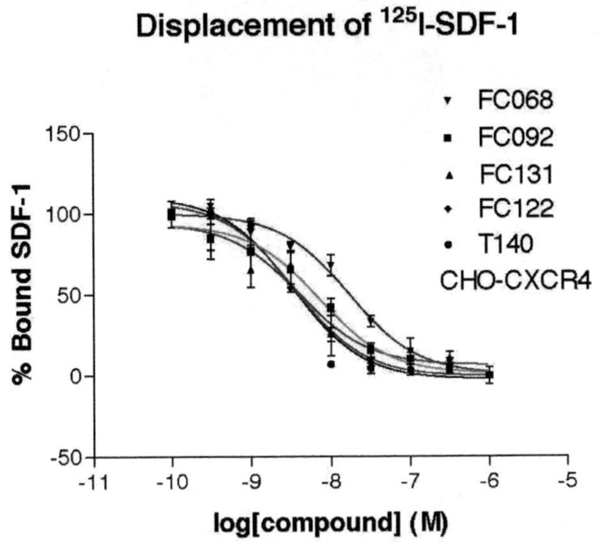

Figure 2

Binding affinity of cyclic pentapeptide inverse agonists for CXCR4. The ability of generation 1-4 cyclic pentapeptide antagonists to inhibit the binding of [^{125}I]CXCL12 to CHO transfectants stably expressing CXCR4 was determined as described in "Materials and methods". The results are representative of three independent binding experiments.

Figure 1

Screening of cyclic pentapeptide candidates for inhibitory activity in Saccharomyces cerevisiae *expressing native CXCR4 stimulated with CXCL12 (A) and the CXCR4-CAM (B) coupled to a pheromone-responsive* FUS1-lacZ *reporter gene. Yeast strains expressing CXCR4 or CXCR4(N119S) were incubated with T140 or cyclic pentapeptide antagonists (10 μM) or vehicle for 6 h at 30°C in a 96 well plate. CXCL12 (2.0 μM) was added to cultures containing yeast expressing native CXCR4 at the same time as the antagonist candidates and incubated for 6 h. Aliquots of each yeast cell culture were solubilized and expression of the pheromone-responsive* lacZ *reporter gene was determined as β-galactosidase enzymatic activity using a fluorogenic substrate. The results were normalized for CXCL12 stimulated activity of native CXCR4 or constitutive activity of the CXCR4-CAM. The relative potencies of the inverse agonist activities of the cyclic pentapeptides on CXCR4-CAM activity is shown in panel C. The results are representative of at least three independent experiments.*

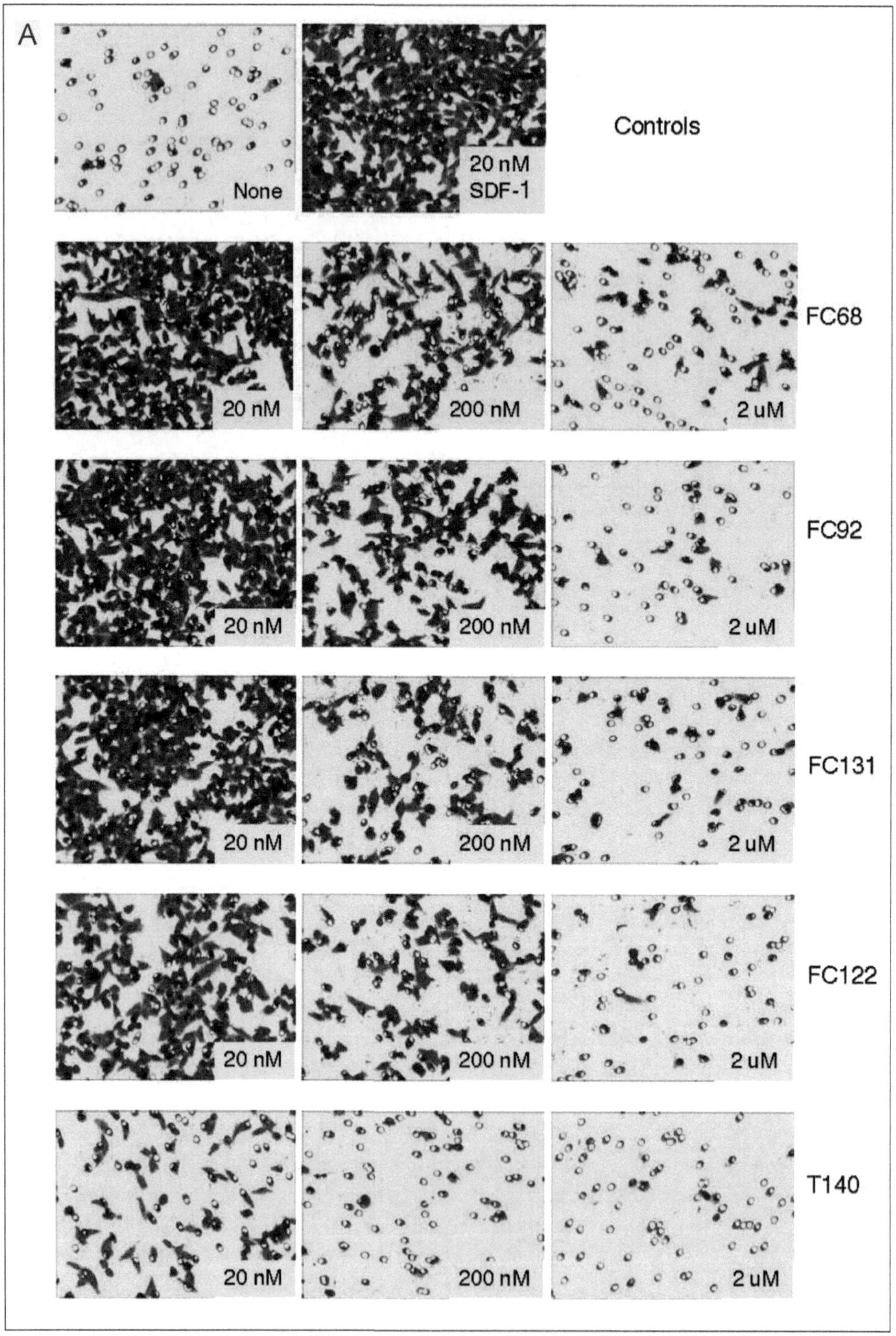
A
None
20 nM SDF-1
Controls
20 nM
200 nM
2 uM
FC68
20 nM
200 nM
2 uM
FC92
20 nM
200 nM
2 uM
FC131
20 nM
200 nM
2 uM
FC122
20 nM
200 nM
2 uM
T140

Characterization of the biologic activity of cyclic pentapeptide CXCR4 antagonists

The potency of the four generations of compounds in blocking the signal transduction induced by CXCL12 binding to CXCR4 was determined in biologic assays. As shown in Figure 3A, the robust chemotactic response of CXCR4 transfectants to 20 nM CXCL12 was abrogated by exposure to 20 nM T140. Similarly, CXCL12-induced chemotaxis was blocked to varying degrees by FC068, FC092, FC131 and FC122. Actual cell counts (four random fields) of the same chemotaxis experiment (Fig. 3B) reveal incremental potency of the sequential generation cyclic pentapeptide compounds when tested at 200 nM. However, differences in the biologic potency are greater than would be expected from the binding studies.

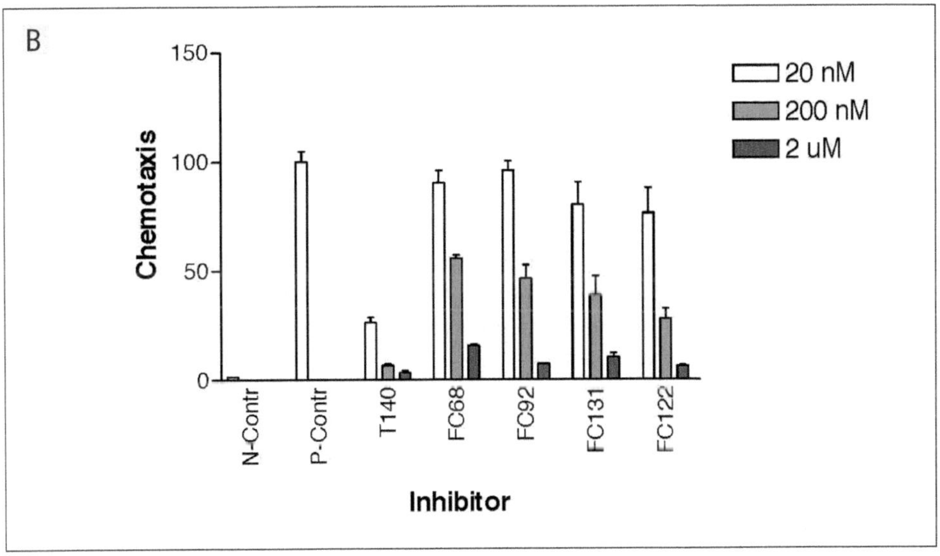

Figure 3
Inhibition of CXCL12 induced chemotaxis of CXCR4 transfectants by cyclic pentapeptide inverse agonists. The directed migration of CXCR4 transfectants to CXCL12 was determined in transwell chambers (panel A) as described in the "Materials and methods" section. Antagonists and CXCL12 were added to the bottom well in the indicated concentrations. Experiments were harvested after 4 h at 37°C. The inserts were washed and stained with crystal violet. Cells were scraped from the upper surface of the membranes, which were then washed and stained with crystal violet. Cells that migrated to the bottom side of the membrane and photographed (panel A) and counted (panel B). The results are representative of three independent experiments.

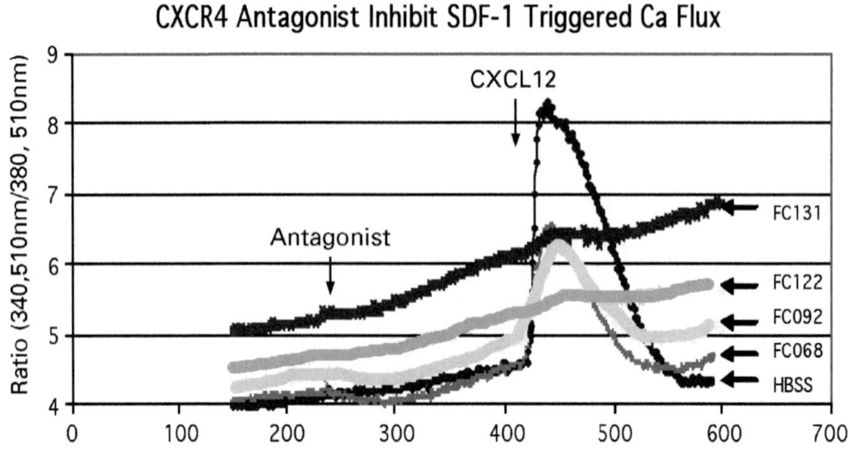

Figure 4
Inhibition of CXCL12 induced signaling of CXCR4 transfectants by cyclic pentapeptide inverse agonists. The inhibition of cytosolic signal transduction by the cyclic pentapeptide antagonists was determined in calcium mobilization experiments as described in "Materials and methods". CHO transfectants stably expressing CXCR4 were loaded with Fura-2 and preincubated with cyclic peptide inverse agonists or buffer controls and tested for a calcium flux response to CXCL12 (1 nM) in a spectrofluorometer. The results are representative of at least three independent experiments.

The efficiency of the four generations of cyclic pentapeptides to blockade the induction of CXCR4 signaling by CXCL12 was also studied in calcium mobilization experiments. Preincubation of CXCR4 transfectants with cyclic pentapeptide antagonists (20 nM) or buffer controls revealed that FC068 and FC092 gave partial inhibition of calcium mobilization induced by 1 nM CXCL12 (Fig. 4). The calcium response to CXCL12 was completely abolished by exposure to FC131 and FC122 (Fig. 4), similar to the effect of T140 (data not shown).

CXCR4-CAM binding affinity correlates with biologic potency of CXCR4 antagonists

The binding properties of the cyclic pentapeptide antagonists demonstrated incremental improvement with each generation and the affinities of FC092, FC131 and

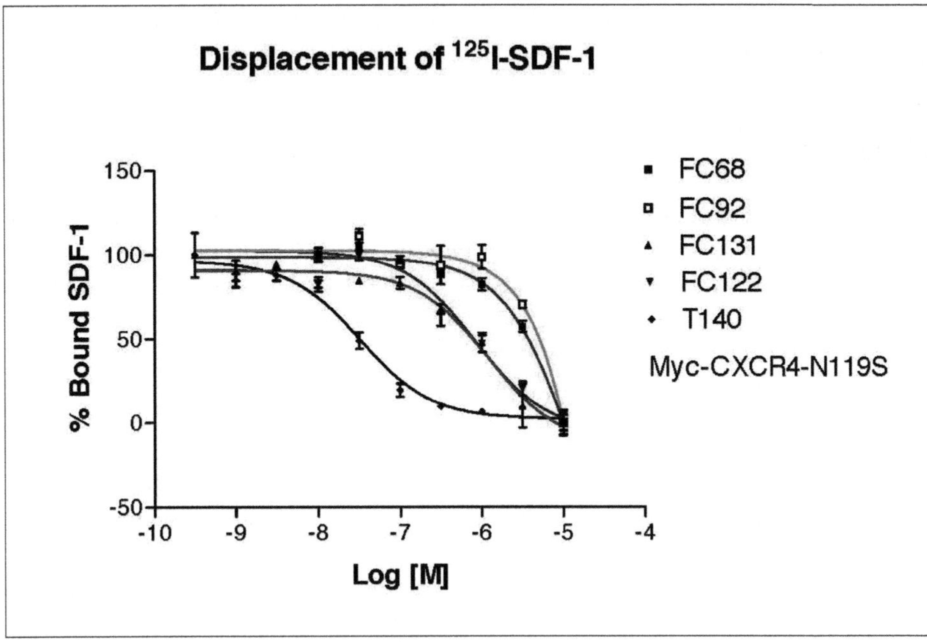

Figure 5
Binding affinity of cyclic pentapeptide inverse agonists for the activated conformation of CXCR4. The ability of cyclic pentapeptide antagonists and the T140 parental template to inhibit the binding of [^{125}I]CXCL12 to CHO transfectants stably expressing CXCR4 was determined as described in "Materials and methods". The results are representative of three independent binding experiments.

FC122 approached that of T140. In contrast, the biologic studies demonstrated significant differences between the ability of the cyclic pentapeptide antagonists and the T140 parental compound to block the ability of CXCL12 to induce cytosolic signaling and program directed migration. Since the ability to reverse the active receptor conformation represents a critical factor in blocking the effect of ligand activation, the binding affinities of the cyclic pentapeptides was determined using the CXCR4-CAM, which reflects an active receptor conformation. As shown in Figure 5, there were significant differences in the ability of these compounds to antagonize the binding of [^{125}I]CXCL12 to transfectants expressing CXCR4(N119S) containing a Myc epitope tag. This modification results in a decrease in affinity of approximately ten-fold. FC068 and FC092 had similar binding affinities (IC_{50} values in the μM range) that were less than those of the third and fourth generation agents, FC131 and FC122 (~1 μM), which were similar. Clearly, T140 was the most potent inhibitor of CXCL12 binding to the CXCR4-CAM. This hierarchy of affinity for the active conformation of CXCR4 closely paralleled that observed in the biologic experiments.

Discussion

Here we report the application of a novel strategy for screening arrays of cyclic pentapeptides that utilizes constitutively active variants of GPCRs to identify antagonists. The GPCR was expressed in yeast cells linked to the pheromone response pathway by a hybrid Gα subunit. The detection system employed a pheromone-responsive *lacZ* reporter gene. The yeast system employed for these experiments is well suited to high throughput screening and the use of constitutively active mutants streamlines the technical aspects of screening and avoids the expense of stimulating native CXCR4 with high concentrations of recombinant CXCL12 required for induction of β-galactosidase activity encoded by the reporter gene. While the ligand activated approach has the potential ability to detect all pharmacologic classes of antagonists, screening with the constitutively active mutants only detects inverse agonists, which shift the active conformation to the inactive state, and weak partial agonists, based on the hypersensitivity of this signaling variant to stimulation. However, in our experience, screening with the native receptor did not have sufficient sensitivity to detect two candidate weak partial agonists that were identified using the constitutively active mutant. While these compounds were not pursued as candidates because they were found to have IC_{50} values greater than 10 μM, the ability of the screening system to detect them provides direct evidence for its exquisite sensitivity. Parallel analysis of this array of compounds using the conventional technique of [^{125}I]CXCL12 displacement failed to identify additional candidates that were not detected using the yeast screening system.

Using the yeast system with human CXCR4 linked to the pheromone response pathway, it was possible to select active cyclic pentapeptides designed from critical amino acid residues from T140, a 14 residue polypeptide derived from tachyplexin, a naturally occurring horseshoe crab protein that was found to inhibit infection by T-tropic strains of HIV-1 over one decade ago, prior to the discovery of HIV-1 coreceptors. The cyclic pentapeptide arrays were composed of four residues identified to be critical for T140 activity in alanine scanning mutants, Napthyl-Ala, Tyr, Arg, and Arg, and a glycine spacer. The library was generated by permuting the position and chirality of the residues in the cyclic pentapeptide format. This strategy enabled the downsizing of T140 activity to approach the essential pharmacophore and come close to the molecular weight of a small molecule. It will be critical to further modify the fourth generation compound, FC122, to reduce the peptide nature in order to attain more favorable pharmacologic properties, including oral bioavailability. Overall, a significant downsizing has been accomplished without significantly compromising the binding affinity for native CXCR4.

The binding affinities of the third and fourth generation compounds, FC131 and FC122, respectively, for native human CXCR4 are virtually identical to that of the parental T140 template, which is a potent antagonist of CXCL12-mediated signaling. The efficacy of these two compounds in inhibiting infection by T-tropic (X4) strains of HIV-1 are also similar to that of T140. In contrast, both were significantly less potent in blocking the biologic activity of CXCR4 induced by CXCL12 stimulation in calcium mobilization and chemotaxis experiments. Thus, although FC131 and FC122 block the binding of CXCL12 to native CXCR4 with similar affinities, their potency in inhibiting signal transduction and directed migration induced by this ligand was approximately an order of magnitude less than that of T140.

Surprisingly, the hierarchy of potency of the four generations of compounds was reflected in their ability to block the binding of CXCL12 to CXCR4(N119S), a constitutively active mutant. We have previously described two constitutively active CXCR4 variants conferred by substitution of Asn-119 with Ser or Ala. CXCR4(N119S) has moderately elevated basal activity in both yeast and mammalian cells that can be further augmented by binding to CXCL12 and interaction with T140 triggers a significant reduction in autonomous signaling. CXCR4(N119A) has a high level of basal activity that is minimally enhanced by exposure to CXCL12 and exposure to T140 results in a limited decrease in constitutive activity. The binding affinities of T140 and the four generations of cyclic pentapeptide antagonists for CXCR4(N119S) provided an accurate insight into the potency of inhibition of biologic outcomes of CXCL12 activation of cells expressing the native receptor. In contrast, the binding affinity of CXCR4(N119A) for all of the cyclic pentapeptides was low and lacked correlation with biologic potency (data not shown). The ability to derive CXCR4-CAMs with different levels of autonomous signaling activity and different binding affinities for cyclic pentapeptide antagonists indicates that multiple activated conformations may be created. The difference in the ability of these two CAMs to predict the biologic potency of the cyclic pentapeptides suggests that CXCR4(N119S) may better represent the active conformation that the native receptor achieves. Thus, while constitutively active GPCR mutants have the potential to facilitate screening and provide key insights into the biologic potency of CXCR4 antagonists, it is critical to select CAMs with the appropriate biochemical properties for this function.

Significant concerns regarding the potential toxicity of CXCR4 antagonists in clinical utilization have been raised because deletion of CXCL12 or CXCR4 in mice is lethal during embryogenesis. The CXCR4 mutations that occur in the WHIM syndrome result in the loss of receptor desensitization [9]. Although these receptors are hyperactivated, the biologic consequence is a partial loss of function. These patients have warts associated with papilloma virus infection, hypogammaglobulinemia, infections, and retention of granulocytes in the bone marrow, but lack the lethal defects in formation of the cerebellum, hematopoietic compartment of bone marrow, and vascular system. The phenotypes of CXCR4/CXCL12 knockout mice [5–7] and patients with the WHIM syndrome [9] reflect the lack of CXCR4 function during embryologic development. The consequences of CXCR4 blockade during adulthood are not clear. In pilot experiments, we have administered second and third generation CXCR4 antagonists to mice in dosages of approximately 2 μg/g (2 mg/kg) every other day for a course of three months. Necropsy did not reveal significant abnormalities and there was no mortality to normal mice. More detailed studies of leukocyte subsets are currently in progress. However, current evidence suggests that administration of CXCR4 antagonists may not have toxicities that preclude clinical utilization.

Two CXCR4 antagonists, AMD3100 [14] and ALX40-4C [13], have already been administered to HIV-1 patients with advanced disease and AMD3100 has been shown to mobilize hematopoietic stem cells in humans [8]. Before the role of CXCR4 blockade in HIV-1 treatment regimens can be accurately evaluated it will be critical to develop antagonists with favorable pharmacokinetic properties. While animal studies suggest that administration of CXCR4 antagonists may block metastatic spread in xenograft models, further clinical development will hinge upon characterization of the consequences of chronic blockade of CXCR4 and, again, the availability of appropriate antagonists. Since CXCL12 is a key factor for programming the directed migration of stem cells and primordial germ cells [4] and CXCR4 antagonists mobilize primitive hematopoietic progenitors [8], short-term administration of these agents may become a critical component of a variety of cellular therapies in the future. The application of the CXCR4-activated pheromone response system in yeast described here should be a valuable tool for developing and characterizing future generations of CXCR4 antagonists.

Acknowledgement

Research described in this article was supported in part by Philip Morris USA Inc. and Philip Morris International. Other support was provided by NIH AI31346. Dr. Peiper's efforts were supported in part by the Georgia Cancer Coalition.

References

1 Bleul CC, Farzan M, Choe H, Parolin C, Clark-Lewis I, Sodroski J, Springer TA (1996) The lymphocyte chemoattractant SDF-1 is a ligand for LESTR/fusin and blocks HIV-1 entry. *Nature* 382: 829–833

2 Oberlin E, Amara A, Bachelerie F, Bessia C, Virelizier JL, Arenzana-Seisdedos F, Schwartz O, Heard JM, Clark-Lewis I, Legler DF et al (1996) The CXC chemokine SDF-1 is the ligand for LESTR/fusin and prevents infection by T-cell-line-adapted HIV-1. *Nature* 382: 833–835

3 Bacon K, Baggiolini M, Broxmeyer H, Horuk R, Lindley I, Mantovani A, Maysushima K, Murphy P, Nomiyama H, Oppenheim J et al; IUIS/WHO Subcommittee on Chemokine Nomenclature (2002) Chemokine/chemokine receptor nomenclature. *J Interferon Cytokine Res* 22: 1067–1068

4 Doitsidou M, Reichman-Fried M, Stebler J, Koprunner M, Dorries J, Meyer D, Esguerra CV, Leung T, Raz E (2002) Guidance of primordial germ cell migration by the chemokine SDF-1. *Cell* 111: 647–659

5 Nagasawa T, Hirota S, Tachibana K, Takakura N, Nishikawa S, Kitamura Y, Yoshida N, Kikutani H, Kishimoto T (1996) Defects of B-cell lymphopoiesis and bone-marrow myelopoiesis in mice lacking the CXC chemokine PBSF/SDF-1. *Nature* 382: 635–638

6 Zou YR, Kottmann AH, Kuroda M, Taniuchi I, Littman DR (1998) Function of the chemokine receptor CXCR4 in haematopoiesis and in cerebellar development. *Nature* 393: 595–599

7 Tachibana K, Hirota S, Iizasa H, Yoshida H, Kawabata K, Kataoka Y, Kitamura Y, Matsushima K, Yoshida N, Nishikawa S et al (1998) The chemokine receptor CXCR4 is essential for vascularization of the gastrointestinal tract. *Nature* 393: 591–594

8 Liles WC, Broxmeyer HE, Rodger E, Wood B, Hubel K, Cooper S, Hangoc G, Bridger GJ, Henson GW, Calandra G, Dale DC (2003) Mobilization of hematopoietic progenitor cells in healthy volunteers by AMD3100, a CXCR4 antagonist. *Blood* 102: 2728–2730

9 Hernandez PA, Gorlin RJ, Lukens JN, Taniuchi S, Bohinjec J, Francois F, Klotman ME, Diaz GA (2003) Mutations in the chemokine receptor gene CXCR4 are associated with WHIM syndrome, a combined immunodeficiency disease. *Nat Genet* 34: 70–74

10 Feng Y, Broder CC, Kennedy PE, Berger EA (1996) HIV-1 entry cofactor: functional cDNA cloning of a seven-transmembrane, G protein-coupled receptor. *Science* 272: 872–877

11 Doranz BJ, Rucker J, Yi Y, Smyth RJ, Samson M, Peiper SC, Parmentier M, Collman RG, Doms RW (1996) A dual-tropic primary HIV-1 isolate that uses fusin and the beta-chemokine receptors CKR-5, CKR-3, and CKR-2b as fusion cofactors. *Cell* 85: 1149–1158

12 Hu QX, Barry AP, Wang ZX, Connolly SM, Peiper SC, Greenberg ML (2000) Evolution of the human immunodeficiency virus type 1 envelope during infection reveals molecular corollaries of specificity for coreceptor utilization and AIDS pathogenesis. *J Virol* 74: 11858–11872

13 Doranz BJ, Filion LG, Diaz-Mitoma F, Sitar DS, Sahai J, Baribaud F, Orsini MJ, Benovic JL, Cameron W, Doms RW (2001) Safe use of the CXCR4 inhibitor ALX40-4C in humans. *AIDS Res Hum Retroviruses* 17: 475–486

14 Hendrix CW, Collier AC Lederman MM, Schols D, Pollard RB, Brown S, Jackson JB, Coombs RW, Glesby MJ, Flexner CW et al; AMD3100 HIV Study Group (2004) Safety, pharmacokinetics, and antiviral activity of AMD3100, a selective CXCR4 receptor inhibitor, in HIV-1 infection. *J Acquir Immune Defic Syndr* 37: 1253–1262

15 Muller A, Homey B, Soto H, Ge N, Catron D, Buchanan ME, McClanahan T, Murphy

E, Yuan W, Wagner SN et al (2001) Involvement of chemokine receptors in breast cancer metastasis. *Nature* 410: 50–56

16 Kang Y, Siegel PM, Shu W, Drobnjak M, Kakonen SM, Cordon-Cardo C, Guise TA, Massague J (2003) A multigenic program mediating breast cancer metastasis to bone. *Cancer Cell* 3: 537–549

17 Smith MC, Luker KE, Garbow JR, Prior JL, Jackson E, Piwnica-Worms D, Luker GD (2004) CXCR4 regulates growth of both primary and metastatic breast cancer. *Cancer Res* 64: 8604–8612

18 Taichman RS, Cooper C, Keller ET, Pienta KJ, Taichman NS, McCauley LK (2002) Use of the stromal cell-derived factor-1/CXCR4 pathway in prostate cancer metastasis to bone. *Cancer Res* 62: 1832–1837

19 Burger M, Glodek A, Hartmann T, Schmitt-Graff A, Silberstein LE, Fujii N, Kipps TJ, Burger JA (2003) Functional expression of CXCR4 (CD184) on small-cell lung cancer cells mediates migration, integrin activation, and adhesion to stromal cells. *Oncogene* 22: 8093–8101

20 Retz MM, Sidhu SS, Blaveri E, Kerr SC, Dolganov GM, Lehmann J, Carroll P, Simko J, Waldman FM, Basbaum C (2005) CXCR4 expression reflects tumor progression and regulates motility of bladder cancer cells. *Int J Cancer* 114: 182–189

21 Castellone MD, Guarino V, De Falco V, Carlomagno F, Basolo F, Faviana P, Kruhoffer M, Orntoft T, Russell JP, Rothstein JL et al (2004) Functional expression of the CXCR4 chemokine receptor is induced by RET/PTC oncogenes and is a common event in human papillary thyroid carcinomas. *Oncogene* 23: 5958–5967

22 Samara GJ, Lawrence DM, Chiarelli CJ, Valentino MD, Lyubsky S, Zucker S, Vaday GG (2004) CXCR4-mediated adhesion and MMP-9 secretion in head and neck squamous cell carcinoma. *Cancer Lett* 214: 231–241

23 Rubin JB, Kung AL, Klein RS, Chan JA, Sun Y, Schmidt K, Kieran MW, Luster AD, Segal RA (2003) A small-molecule antagonist of CXCR4 inhibits intracranial growth of primary brain tumors. *Proc Natl Acad Sci USA* 100: 13513–13518

24 Geminder H, Sagi-Assif O, Goldberg L, Meshel T, Rechavi G, Witz IP, Ben-Baruch A (2001) A possible role for CXCR4 and its ligand, the CXC chemokine stromal cell-derived factor-1, in the development of bone marrow metastases in neuroblastoma. *J Immunol* 167: 4747–4757

25 Jankowski K, Kucia M, Wysoczynski M, Reca R, Zhao D, Trzyna E, Trent J, Peiper S, Zembala M, Ratajczak J et al (2003) Both hepatocyte growth factor (HGF) and stromal-derived factor-1 regulate the metastatic behavior of human rhabdomyosarcoma cells, but only HGF enhances their resistance to radiochemotherapy. *Cancer Res* 63: 7926–7935

26 Burger JA, Burger M, Kipps TJ (1999) Chronic lymphocytic leukemia B cells express functional CXCR4 chemokine receptors that mediate spontaneous migration beneath bone marrow stromal cells. *Blood* 94: 3658–3667

27 Burger JA, Spoo A, Dwenger A, Burger M, Behringer D (2003) CXCR4 chemokine receptors (CD184) and alpha4beta1 integrins mediate spontaneous migration of human

CD34+ progenitors and acute myeloid leukaemia cells beneath marrow stromal cells (pseudoemperipolesis). *Br J Haematol* 122: 579–589

28 Liang Z, Wu T, Lou H, Yu X, Taichman RS, Lau SK, Nie S, Umbreit J, Shim H (2004) Inhibition of breast cancer metastasis by selective synthetic polypeptide against CXCR4. *Cancer Res* 64: 4302–4308

29 Ichiyama K, Yokoyama-Kumakura S, Tanaka Y, Tanaka R, Hirose K, Bannai K, Edamatsu T, Yanaka M, Niitani Y, Miyano-Kurosaki N et al (2003) A duodenally absorbable CXC chemokine receptor 4 antagonist, KRH-1636, exhibits a potent and selective anti-HIV-1 activity. *Proc Natl Acad Sci USA* 100: 4185–4190

30 Fujii N, Oishi S, Hiramatsu K, Araki T, Ueda S, Tamamura H, Otaka A, Kusano S, Terakubo S, Nakashima H et al (2003) Molecular-size reduction of a potent CXCR4-chemokine antagonist using orthogonal combination of conformation- and sequence-based libraries. *Angew Chem Int Ed Engl* 42: 3251–3253

31 Zhang WB, Navenot JM, Haribabu B, Tamamura H, Hiramatu K, Omagari A, Pei G, Manfredi JP, Fujii N, Broach JR, Peiper SC (2002) A point mutation that confers constitutive activity to CXCR4 reveals that T140 is an inverse agonist and that AMD3100 and ALX40-4C are weak partial agonists. *J Biol Chem* 277: 24515–24521

Antagonists of CXCR3: a review of current progress

Tassie L. Collins[1], Michael G. Johnson[2], Julio C. Medina[2]

[1]Department of Neuroscience; [2]Department of Chemistry, Amgen Inc., 1120 Veterans Boulevard, South San Francisco, CA 94080, USA

Introduction

CXCR3 was cloned and identified as a receptor for CXCL9 and CXCL10 by Loetscher, et al., in 1996 [1], and was subsequently identified as a receptor for CXCL11 [2]. CXCL9, CXCL10 and CXCL11 are selective, potent agonists of CXCR3 (Kd 0.1–5 nM) [1–4]. Additional chemokines have been reported to bind to CXCR3 (e.g., CXCL13 [5] and CCL11 [6]), however the reported affinities are generally weak and the biological significance of the interactions is questionable. Similarly, the ligands for CXCR3 have been reported to be antagonists of CCR3 [7] and CCR5 [8], but high concentrations of the CXCR3 ligands are required to achieve inhibition of CCR3 or CCR5 biological functions.

CXCR3 and its ligands have been implicated in a large variety of immune disorders, including organ transplant rejection and multiple sclerosis. Cells expressing CXCR3 are found at high numbers in biopsies taken from patients experiencing organ transplant rejection [9–17]. CXCL9 [11–13, 15, 16], CXCL10 [9–13, 15–17] and CXCL11 [11–17] message and protein levels are increased in the tissues of organs undergoing rejection. Importantly, the levels of CXCR3$^+$ cells and CXCR3 ligand mRNA in the biopsy samples correlate with the grade of graft rejection [10, 12–14, 16], suggesting a causative role for CXCR3 in transplant rejection. This is supported by transplant studies performed in rodents. Neutralizing antibodies to CXCL9 [20], CXCL10 [19] or to CXCR3 [18] prolong cardiac allograft survival. Mice lacking expression of CXCR3 tolerate a cardiac allograft significantly longer than wild-type mice [18]. In the presence of sub-therapeutic doses of cyclosporine A, CXCR3-deficient mice will permanently accept a cardiac allograft [18]. In reciprocal studies, cardiac allografts taken from CXCL10-deficient mice show prolonged survival in wild-type mice, demonstrating that production of CXCL10 by the cardiac tissues is a critical component of transplant rejection [19]. Blockade of the CXCR3 pathway has also been shown to be effective in experimental models of lung allograft [21, 24], small bowel allograft [25], islet allograft [22] and graft-versus-host disease [23, 26].

Chemokine Biology – Basic Research and Clinical Application, Volume II
edited by Kuldeep Neote, Gordon L. Letts and Bernhard Moser

A large body of research also suggests a key role for the CXCR3 pathway in multiple sclerosis. In patients with multiple sclerosis, 80% of $CD4^+$ T cells and 97% of $CD8^+$ cells in the cerebrospinal fluid (CSF) express CXCR3 [27]. Increased concentrations of CXCL9 and CXCL10 are found in brain lesions of patients with multiple sclerosis [28–30]. Furthermore, elevated levels of CXCL10 are found in the CSF of patients experiencing a relapse in disease [27–30]. Notably, there is a significant correlation between the levels of CXCL10 and the number of T cells in the CSF [27]. In addition, there is a significant correlation between the levels of CXCL10 in the CSF and the time to disease relapse [30]. $CXCR3^+$ T cells are also found in brain lesions of patients with multiple sclerosis [27–31] and it appears that these cells continuously accumulate in the lesions, as the number of $CXCR3^+$ cells increases with the age of the lesion [31]. Collectively, these data suggest that the CXCR3 pathway plays a causative role in the pathology of multiple sclerosis. In animal models, mixed results have been reported for this pathway. CXCL10-deficient mice have exacerbated disease in the active immunization model of EAE, as do rats treated with a neutralizing antibody to CXCL10 [32, 33]. However, neutralizing antibody to CXCL10 appears to ameliorate disease in the passive immunization ('adoptive transfer') model of EAE [34]. Moreover, in a viral-induced model of neuroinflammation, CXCL10-deficient mice have decreased numbers of $CD4^+$ and $CD8^+$ T cells in the brain correlating with diminished demyelination [35]. In this model, neutralizing antibodies to CXCL10 also diminish T cell trafficking to the brain and decrease demyelination [36]. It is uncertain whether the differing results stem from the use of different models or from the use of different reagents. It is worth noting that animal models for multiple sclerosis are particularly poor at predicting success in human disease [37, 38], so that the relevance of the animal data for CXCR3 is somewhat questionable.

The available data suggest that agents that antagonize the CXCR3 pathway could be useful in the treatment of human inflammatory and immune diseases. In this review, we will report on the current progress toward identifying agents that antagonize the CXCR3 pathway. The approaches that have been taken toward this end include a number of modalities: small molecule antagonists of CXCR3, neutralizing antibodies to CXCR3 or CXCL10, and antagonistic modifications of CXCL10 and CXCL11.

While much of the research that provides the biological rationale for targeting the CXCR3 pathway has been performed by academic researchers and has been reported in peer-reviewed journals, most of the effort aimed at antagonizing the CXCR3 pathway is being performed by researchers in the pharmaceutical industry. Little of this work has yet been presented in peer-reviewed journals. Most of the available information is found by searching the patent literature. In order to provide a comprehensive overview, we have included this information in our review of the field.

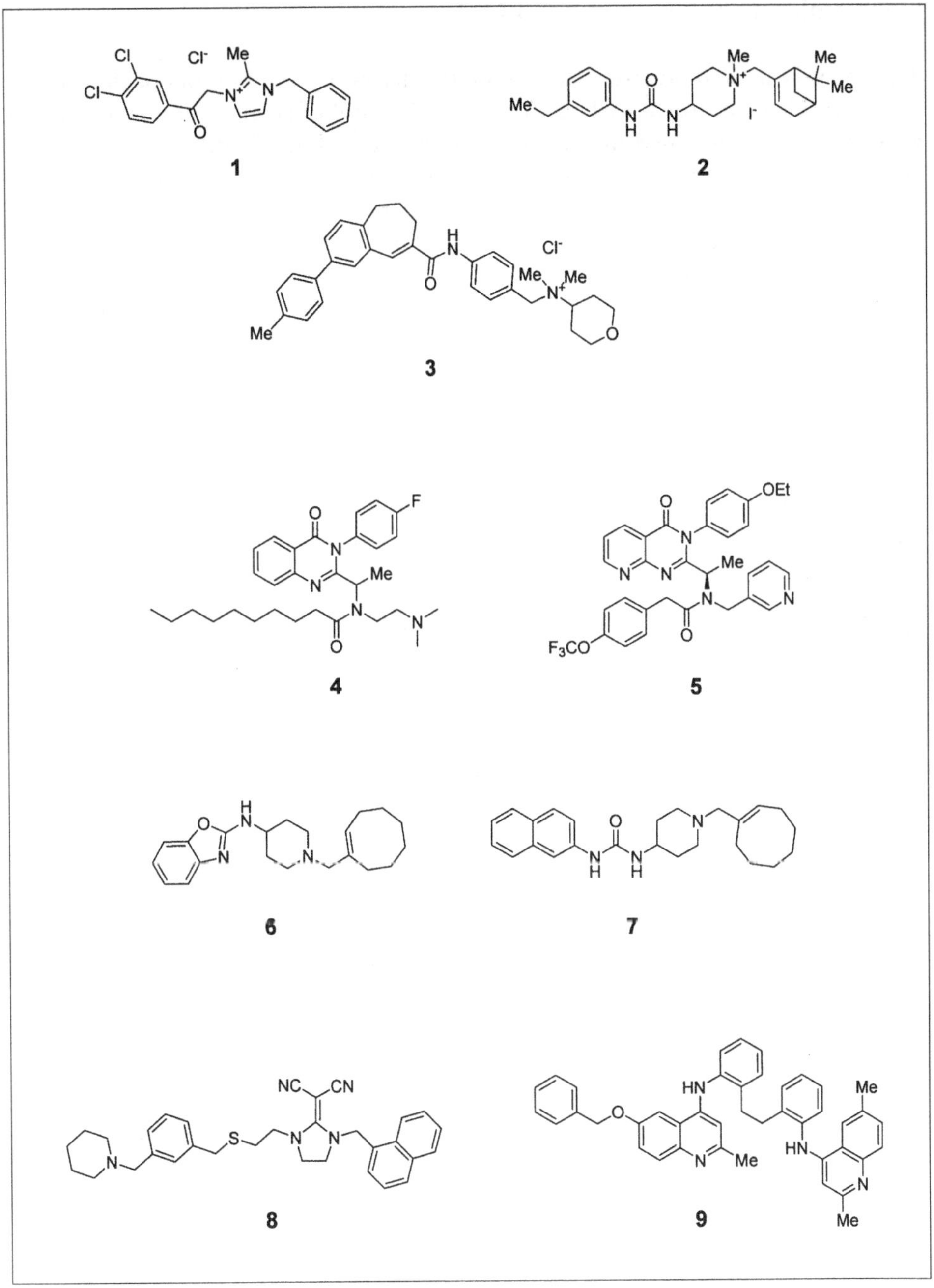

Figure 1
Small molecule antagonists of CXCR3

Small molecule antagonists of CXCR3

There is very limited literature describing the discovery of inhibitors of the chemotactic cytokine receptor CXCR3. However, a handful of small molecule antagonist series have been disclosed in the patent literature and at scientific meetings. As a class, inventors describe the small molecule CXCR3 antagonists as potential treatments for a broad range of autoimmune disorders.

In general, there is no obvious structural motif that all of the antagonist series possess. However, several of the antagonist series share a quaternary ammonium salt functionality such as an imidazolium (**1**), a piperidinium (**2**) or a dimethylaminium moiety (**3**) [39–41]. Currently, there is no inhibitor-receptor model that rationalizes the binding affinities of any of the CXCR3 antagonists to the receptor. Ranking of binding affinities between antagonist series is difficult because of the different assays used to evaluate the antagonist activities and the general lack of data revealed in the patent literature.

While most CXCR3 antagonists described in the literature claim to be CXCR3 selective, TAK-779 (**3**) is described as a CCR5 and CXCR3 antagonist [41]. This compound was first identified as a CCR5 antagonist for the potential treatment of HIV. In 2002, Gao et al. reported that TAK-779 inhibits binding and functional activity of cells transfected with the mouse CXCR3 receptor.

A series of quinazolinone [42] and 8-azaquinazolinone [43] derivatives have been reported in the patent literature as inhibitors of CXCR3. The most advanced of these compounds is T0906487 (**5**), which has been evaluated in a phase 2a psoriasis trial [44]. T0906487 has been reported to inhibit binding of [^{125}I]-labeled CXCL10 and CXCL11 with IC_{50} values of 7.4 nM and 8.2 nM, respectively. This compound is also reported to inhibit cell migration *in vitro* mediated by the three known CXCR3 chemokines, CXCL9, CXCL10 and CXCL11 with IC_{50} values ranging from 8–36 nM [45]. Phase 1 clinical studies indicated the compound is well-tolerated and safe for both single and multiple dosing [46].

A series of piperidine derivatives exemplified by compounds **2**, **6** and **7**, have been disclosed as CXCR3 antagonists [47, 48]. Even though no specific values were reported, these compounds were evaluated in a FLIPR assay and are claimed to inhibit CXCL9, CXCL10 and CXCL11 binding to CXCR3.

Imidazolidine and aminoquinoline derivatives, represented by **8** and **9** respectively, represent another group of molecules reported in the patent literature to be inhibitors of CXCR3 [49, 50]. Analogs of the imidazolidine series are reported to inhibit CXCL10 binding to CXCR3 in a radioligand binding assay.

Neutralizing antibodies

Neutralizing antibodies directed against CXCR3, CXCL9 and CXCL10 have been shown to be efficacious in animal models of disease including neuroinflammation [34, 36], arthritis [51], inflammatory bowel disease [52], graft-*versus*-host disease [26], and transplant rejection [18–22]. Although a variety of researchers have reported data using antibodies developed in rodent species, there have as yet been no reports documenting the development of therapeutic antibodies. However, the patent literature details some of the approaches that are being tried.

Howard et al. and Arimilli et al. report on ideal CXCR3-derived antigens for generating neutralizing antibodies to CXCR3 [53, 54]. They describe the use of cyclized peptides derived from the first and/or fourth extracellular domains of CXCR3. Data is provided in their patent filings for antibodies derived using the sequence 'SP-4-1', which contains amino acids from both the fourth and first extracellular domain. Immunization of SJL mice with the SP-4-1 peptide at weeks 3 and 2 prior to immunization with the encephalitogenic peptide of proteolipid protein ($PLP_{139-151}$) results in generation of antibodies directed against CXCR3 and these animals experience decreased disease severity and incidence [53, 54]. They additionally report that treatment of SJL mice with a monoclonal antibody generated against a 37-amino acid peptide derived from the first extracellular domain of CXCR3 on days –1, +3 and +7 of immunization with $PLP_{139-151}$ results in decreased disease incidence. The inventors claim use of these peptides to generate a monoclonal antibody for therapeutic use [54] or for generating an autoantibody response directed against CXCR3 [53] for treatment of multiple sclerosis.

Balasa et al. [55] report on antibodies that bind to CXCL10, most particularly for the treatment of inflammatory bowel diseases. The antibodies described in the patent application are neutralizing antibodies directed against human CXCL10 and have low nanomolar to high picomolar potency for CXCL10.

The current development status of these antibodies is not known.

Modified ligands

It is known from the literature that modification of the ligands for CXCR3 can convert them from potent agonists into potent antagonists [56, 57]. Modification of CXCL9, CXCL10 and CXCL11 may occur naturally, since these chemokines undergo amino-terminal cleavage in the presence of CD26/dipeptidyl peptidase IV [57]. These N-terminally truncated chemokines fail to induce calcium flux or chemotaxis and have reduced receptor-binding affinity. However, preincubation of CXCR3-expressing cells with the truncated chemokines reduces subsequent calcium mobilization in response to intact ligands, indicating that the truncated chemokines can bind to and antagonize CXCR3.This is further supported by the finding that the truncated chemokines will inhibit migration to intact ligands [57]. Interestingly, despite the loss of CXCR3 agonistic properties, N-terminally truncated CXCL9 and CXCL10 retained their angiostatic properties [57]. N-terminally truncated CXCL9,

CXCL10 and CXCL11 are claimed in a patent application for the treatment of inflammatory diseases and diseases in which angiogenesis is thought to play an important role [58]. No data are available on the *in vivo* efficacy of the N-terminal truncated CXCR3 ligands in animal models for disease.

It has been demonstrated for several chemokines (CCL2, CCL4, CCL5) that the ability to bind to glycosaminoglycans and to form oligomeric complexes is crucial to their ability to recruit cells *in vivo* [59, 60]. Proudfoot et al. [61] describe C-terminal modifications (mutation of certain basic residues) of CXCL11 which result in reduced binding to glycosaminoglycans. As reported for CCL2, CCL4 and CCL5 [59], CXCL11 mutants that have reduced glycosaminoglycan binding may retain the ability to stimulate chemotaxis *in vitro* but lose the ability to recruit cells *in vivo*. Interestingly, such mutants may function as antagonists of CXCR3 *in vivo*, as they can inhibit recruitment of cells to the peritoneum by wild-type CXCL11 [61].

Conclusion

CXCR3 is a novel target and it remains to be seen whether agents that antagonize this pathway will be efficacious in the treatment of human disease. Nonetheless, there is great interest in the pharmaceutical industry in agents that can block leukocyte trafficking as evidenced by the number of companies and research groups that are working in the field. The next few years hold a lot of promise as the various approaches described above mature, hopefully resulting in the development of therapeutic agents that advance into clinical trials. It is the expectation that as the field matures, some of these CXCR3 antagonists will also help further elucidate the role that this receptor plays in mediating immune disorders.

References

1 Loetscher M, Gerber B, Loetscher P, Jones SA, Piali L, Clark-Lewis I, Baggiolini M, Moser B (1996) Chemokine receptor specific for IP10 and mig: structure, function, and expression in activated T-lymphocytes. *J Exp Med* 184: 799–802

2 Cole KE, Strick CA, Paradis TJ, Ogborne KT, Loetscher M, Gladue RP, Lin W, Boyd JG, Moser B, Wood DE et al (1998) Interferon-inducible T cell alpha chemoattractant (I-TAC): a novel non-ELR CXC chemokine with potent activity on activated T cells through selective high affinity binding to CXCR3. *J Exp Med* 187: 2009–2021

3 Weng Y, Siciliano SJ, Waldburger KE, Sirotina-Meisher A, Staruch MJ, Daugherty BL, Gould SL, Springer MS, DeMartino JA (1998) Binding and functional properties of recombinant and endogenous CXCR3 chemokine receptors. *J Biol Chem* 273: 18288–18291

4 Loetscher M, Loetscher P, Brass N, Meese E, Moser B (1998) Lymphocyte-specific

chemokine receptor CXCR3: regulation, chemokine binding and gene localization. *Eur J Immunol* 28: 3696–3705

5 Jenh CH, Cox MA, Hipkin W, Lu T, Pugliese-Sivo C, Gonsiorek W, Chou CC, Narula SK, Zavodny PJ (2001) Human B cell-attracting chemokine 1 (BCA-1; CXCL13) is an agonist for the human CXCR3 receptor. *Cytokine* 15: 113–121

6 Xanthou G, Duchesnes CE, Williams TJ, Pease JE (2003) CCR3 functional responses are regulated by both CXCR3 and its ligands CXCL9, CXCL10 and CXCL11. *Eur J Immunol* 33: 2241–2250

7 Loetscher P, Pellegrino A, Gong JH, Mattioli I, Loetscher M, Bardi G, Baggiolini M, Clark-Lewis I (2001) The ligands of CXC chemokine receptor 3, I-TAC, Mig, and IP10, are natural antagonists for CCR3. *J Biol Chem* 276: 2986–2991

8 Petkovic V, Moghini C, Paoletti S, Uguccioni M, Gerber B (2004) I-TAC/CXCL11 is a natural antagonist for CCR5. *J Leukoc Biol* 76: 701–708

9 Agostini C, Calabrese F, Rea F, Facco M, Tosoni A, Loy M, Binotto G, Valente M, Trentin L, Semenzato G (2001) Cxcr3 and its ligand CXCL10 are expressed by inflammatory cells infiltrating lung allografts and mediate chemotaxis of T cells at sites of rejection. *Am J Pathol* 158: 1703–1711

10 Melter M, Exeni A, Reinders ME, Fang JC, McMahon G, Ganz P, Hancock WW, Briscoe DM (2001) Expression of the chemokine receptor CXCR3 and its ligand IP-10 during human cardiac allograft rejection. *Circulation* 104: 2558–2564

11 Goddard S, Williams A, Morland C, Qin S, Gladue R, Hubscher SG, Adams DH (2001) Differential expression of chemokines and chemokine receptors shapes the inflammatory response in rejecting human liver transplants. *Transplantation* 72: 1957–1967

12 Zhao DX, Hu Y, Miller GG, Luster AD, Mitchell RN, Libby P (2002) Differential expression of the IFN-gamma-inducible CXCR3-binding chemokines, IFN-inducible protein 10, monokine induced by IFN, and IFN-inducible T cell alpha chemoattractant in human cardiac allografts: association with cardiac allograft vasculopathy and acute rejection. *J Immunol* 169: 1556–1560

13 Fahmy NM, Yamani MH, Starling RC, Ratliff NB, Young JB, McCarthy PM, Feng J, Novick AC, Fairchild RL (2003) Chemokine and chemokine receptor gene expression indicates acute rejection of human cardiac transplants. *Transplantation* 75: 72–78

14 Kao J, Kobashigawa J, Fishbein MC, MacLellan WR, Burdick MD, Belperio JA, Strieter RM (2003) Elevated serum levels of the CXCR3 chemokine ITAC are associated with the development of transplant coronary artery disease. *Circulation* 107: 1958–1961

15 Fahmy NM, Yamani MH, Starling RC, Ratliff NB, Young JB, McCarthy PM, Feng J, Novick AC, Fairchild RL (2003) Chemokine and receptor-gene expression during early and late acute rejection episodes in human cardiac allografts. *Transplantation* 75: 2044–2047

16 Hu H, Aizenstein BD, Puchalski A, Burmania JA, Hamawy MM, Knechtle SJ (2004) Elevation of CXCR3-binding chemokines in urine indicates acute renal-allograft dysfunction. *Am J Transplant* 4: 432–437

17 Panzer U, Reinking RR, Steinmetz OM, Zahner G, Sudbeck U, Fehr S, Pfalzer B, Schnei-

der A, Thaiss F, Mack M et al (2004) CXCR3 and CCR5 positive T-cell recruitment in acute human renal allograft rejection. *Transplantation* 78: 1341–1350

18 Hancock WW, Lu B, Gao W, Csizmadia V, Faia K, King JA, Smiley ST, Ling M, Gerard NP, Gerard C (2000) Requirement of the chemokine receptor CXCR3 for acute allograft rejection. *J Exp Med* 192: 1515–1520

19 Hancock WW, Gao W, Csizmadia V, Faia KL, Shemmeri N, Luster AD (2001) Donor-derived IP-10 initiates development of acute allograft rejection. *J Exp Med* 193: 975–980

20 Miura M, Morita K, Kobayashi H, Hamilton TA, Burdick MD, Strieter RM, Fairchild RL (2001) Monokine induced by IFN-gamma is a dominant factor directing T cells into murine cardiac allografts during acute rejection. *J Immunol* 167: 3494–3504

21 Belperio JA, Keane MP, Burdick MD, Lynch JP 3rd, Xue YY, Li K, Ross DJ, Strieter RM (2002) Critical role for CXCR3 chemokine biology in the pathogenesis of bronchiolitis obliterans syndrome. *J Immunol* 169: 1037–1049

22 Baker MS, Chen X, Rotramel AR, Nelson JJ, Lu B, Gerard C, Kanwar Y, Kaufman DB (2003) Genetic deletion of chemokine receptor CXCR3 or antibody blockade of its ligand IP-10 modulates posttransplantation graft-site lymphocytic infiltrates and prolongs functional graft survival in pancreatic islet allograft recipients. *Surgery* 134: 126–133

23 Duffner U, Lu B, Hildebrandt GC, Teshima T, Williams DL, Reddy P, Ordemann R, Clouthier SG, Lowler K, Liu C et al (2003) Role of CXCR3-induced donor T-cell migration in acute GVHD. *Exp Hematol* 31: 897–902

24 Belperio JA, Keane MP, Burdick MD, Lynch JP 3rd, Zisman DA, Xue YY, Li K, Ardehali A, Ross DJ, Strieter RM (2003) Role of CXCL9/CXCR3 chemokine biology during pathogenesis of acute lung allograft rejection. *J Immunol* 171: 4844–4852

25 Zhang Z, Kaptanoglu L, Tang Y, Ivancic D, Rao SM, Luster A, Barrett TA, Fryer J (2004) IP-10-induced recruitment of CXCR3 host T cells is required for small bowel allograft rejection. *Gastroenterology* 126: 809–818

26 Hildebrandt GC, Corrion LA, Olkiewicz KM, Lu B, Lowler K, Duffner UA, Moore BB, Kuziel WA, Liu C, Cooke KR (2004) Blockade of CXCR3 receptor:ligand interactions reduces leukocyte recruitment to the lung and the severity of experimental idiopathic pneumonia syndrome. *J Immunol* 173: 2050–2059

27 Sorensen TL, Tani M, Jensen J, Pierce V, Lucchinetti C, Folcik VA, Qin S, Rottman J, Sellebjerg F, Strieter RM et al (1999) Expression of specific chemokines and chemokine receptors in the central nervous system of multiple sclerosis patients. *J Clin Invest* 103: 807–815

28 Balashov KE, Rottman JB, Weiner HL, Hancock WW (1999) CCR5(+) and CXCR3(+) T cells are increased in multiple sclerosis and their ligands MIP-1alpha and IP-10 are expressed in demyelinating brain lesions. *Proc Natl Acad Sci* 96: 6873–6878

29 Simpson JE, Newcombe J, Cuzner ML, Woodroofe MN (2000) Expression of the interferon-gamma-inducible chemokines IP-10 and Mig and their receptor, CXCR3, in multiple sclerosis lesions. *Neuropathol Appl Neurobiol* 26: 133–142

30 Mahad DJ, Howell SJ, Woodroofe MN (2002) Expression of chemokines in the CSF and

correlation with clinical disease activity in patients with multiple sclerosis. *J Neurol Neurosurg Psychiatry* 72: 498–502
31 Sorensen TL, Trebst C, Kivisakk P, Klaege KL, Majmudar A, Ravid R, Lassmann H, Olsen DB, Strieter RM, Ransohoff RM et al (2002) Multiple sclerosis: a study of CXCL10 and CXCR3 co-localization in the inflamed central nervous system. *J Neuroimmunol* 127: 59–68
32 Klein RS, Izikson L, Means T, Gibson HD, Lin E, Sobel RA, Weiner HL, Luster AD (2004) IFN-inducible protein 10/CXC chemokine ligand 10-independent induction of experimental autoimmune encephalomyelitis. *J Immunol* 172: 550–559
33 Narumi S, Kaburaki T, Yoneyama H, Iwamura H, Kobayashi Y, Matsushima K (2002) Neutralization of IFN-inducible protein 10/CXCL10 exacerbates experimental autoimmune encephalomyelitis. *Eur J Immunol* 32: 1784–1791
34 Fife BT, Kennedy KJ, Paniagua MC, Lukacs NW, Kunkel SL, Luster AD, Karpus WJ (2001) CXCL10 (IFN-γ-inducible protein-10) control of encephalitogenic $CD4^+$ T cell accumulation in the central nervous system during experimental autoimmune encephalomyelitis. *J Immunol* 166: 7617–7624
35 Dufour JH, Dziejman M, Liu MT, Leung JH, Lane TE, Luster AD (2002) IFN-g-inducible protein 10 (IP-10; CXCL10)-deficient mice reveal a role for IP-10 in effector T cell generation and trafficking. *J Immunol* 168: 3195–3204
36 Liu MT, Chen BP, Oertel P, Buchmeier MJ, Armstrong D, Hamilton TA, Lane TE (2000) The T cell chemoattractant IFN-inducible protein 10 is essential in host defense against viral-induced neurologic disease. *J Immunol* 165: 2327–2330
37 Panitch HS, Hirsch RL, Haley AS, Johnson KP (1987) Exacerbations of multiple sclerosis in patients treated with gamma interferon. *Lancet* 1(8538): 893–895
38 Skurkovich S, Boiko A, Beliaeva I, Buglak A, Alekseeva T, Smirnova N, Kulakova O, Tchechonin V, Gurova O, Deomina T et al (2001) Randomized study of antibodies to IFN-gamma and TNF-alpha in secondary progressive multiple sclerosis. *Multiple Sclerosis* 7: 277–284
39 Axte JM, Foley JJ, Kingsbury WD, Sarau HM (2003) International publication WO 03/101970
40 Watson RJ, Meissner JWG, Owen DA (2004) International publication WO 04/94381
41 Gao P, Zhou XY, Yashiro-Ohtani Y, Yang YF, Sugimoto N, Ono S, Nakanishi T, Obikata S, Imanishi T, Egawa T et al (2003) The unique target specificity of a nonpeptide chemokine receptor antagonist: selective blockade of two TH1 chemokine receptors CCR5 and CXCR3. *J Leukocyte Biol* 73: 273–280
42 Schall TJ, Dairaghi DJ, McMaster B (2001) International publication WO 01/16114
43 Medina JC, Johnson MG, Li AR, Jiwen L, Huang AX, Zhu L, Marcus AP (2002) International publication WO 02/083143
44 Berry K, Friedrich M, Kersey K, Stempien MJ, Wagner F, van Lier JJ, Sabat R, Wolk K (2004) Evaluation of T0906487, a CXCR3 antagonist, in a Phase 2a Psoriasis Trial. Inflammation Research Association Biannual meeting, Lake George, NY, September, 2004. *Inflammation Research* 53: S222

45 Collins TL, Johnson M, Li A.-R, Liu J, Huang A, Zhu L, Marcus A, Danao J, Sablan E, Kumer J et al (2003) T487: A selective and potent small molecule antagonist of CXCR3. 6th World Congress on Inflammation, Vancouver, British Columbia, Canada; August 02–06 2003. *Inflammation Research* 52 (Suppl 2): 118

46 Floren LC, Berry K, Tonn G, Ye Q, Wright M, Huang AX, Wang X, Marcus A, Johnson M, Collins T et al (2003) T0906487 (T487), a novel CXCR3 antagonist: First time in human study of safety and pharmacokinetics. 6th World Congress on Inflammation, Vancouver, British Columbia, Canada; August 02–06 2003. *Inflammation Research* 52 (Suppl 2): 159

47 Owen DA, Watson RJ, Meissner JWG, Allen DR (2005) International publication WO 05/003127

48 Watson RJ, Meissner JWG, Christie MI, Owen DA (2003) International publication WO 03/070242

49 Ohshima E, Sone H, Kotera O, Komatsu R, Larosa GJ, Luly JR (2002) International publication WO 02/085862

50 Lin C-C, Liu J-F, Chang L-W, Chen S-J, Xiang Y, Cheng P, Jang J-J (2004) US Patent application publication US 2004/0209902

51 Salomon I, Netzer N, Wildbaum G, Schif-Zuck S, Maor G, Karin N (2002) Targeting the function of IFN-γ-inducible protein 10 suppresses ongoing adjuvant arthritis. *J Immunol* 169: 2685–2693

52 Singh UP, Singh S, Taub DD, Lillard JW Jr. (2003) Inhibition of IFN-gamma-inducible protein-10 abrogates colitis in IL-$10^{-/-}$ mice. *J Immunol* 171: 1401–1406

53 Howard M, Deshpande S, Ferlin W, Arimilli S. US patent 6,171, 590

54 Arimilli S, Ferlin W, Deshpande S, Mocci S (2002) US patent application publication 2002/0039578

55 Balasa B, Tsurushita N, Landolfi N (2004) International publication WO 04/101511

56 Clark-Lewis I, Mattioli I, Gong J-H, Loetscher P (2003) Structure-function relationship between the human chemokine receptor CXCR3 and its ligands. *J Biol Chem* 278: 289–295

57 Proost P, Schutyser E, Menten P, Struyf S, Wuyts A, Opdenakker G, Detheux M, Parmentier M, Durinx C, Lambeir A-M et al (2001) Amino-terminal truncation of CXCR3 agonists impairs receptor signaling and lymphocyte chemotaxis, while preserving antiangiogenic properties. *Blood* 98: 3554–3561

58 Van Damme J, Proost P (2002) International publication WO 02/059301

59 Proudfoot AE, Handel TM, Johnson Z, Lau EK, LiWang P, Clark-Lewis I, Borlat F, Wells TNC, Kosco-Vilbois MH (2003) Glycosaminoglycan binding and oligomerization are essential for the *in vivo* activity of certain chemokines. *Proc Natl Acad Sci* 100: 1885–1890

60 Lau EK, Paavola CD, Johnson Z, Gaudry J-P, Geretti E, Borlat F, Kungl AJ, Proudfoot AE, Handel TM (2004) Identification of the glycosaminoglycan binding site of the CC chemokine, MCP-1. *J Biol Chem* 279: 22294–22305

61 Proudfoot A, Kosco-Vilbois M (2003) International publication WO 03/106488

IL-8 receptor antagonist: basic research and clinical utility

John R. White[1] *and Henry M. Sarau*[2]

[1]Biopharmaceutical CEDD and [2]Respiratory and Inflammation CEDD, GlaxoSmithKline, P.O. Box 1539, King of Prussia, PA 19406, USA

Introduction

Inflammatory cells are thought to be instrumental in the pathophysiology of diseases and the control of their recruitment and activation appears to be an attractive strategy for therapeutic intervention. Chemokines are a family of small molecular weight (7–15 kDa) proteins that in conjunction with adhesion molecules play a crucial role in leukocyte recruitment, cellular activation and proliferation at sites of inflammation. Chemokines are produced by a variety of cell types, including leukocytic and non-leukocytic cells, usually in response to antigens, irritants and other cytokines. Interleukin-8 (CXCL8) was the first member to be identified of this new family of proinflammatory chemokines that now constitute over 45 members. Chemokines produce their biological effects by interacting with greater than 18 G protein coupled cell surface receptors. A few chemokines bind selectively to a single receptor but other chemokines bind to more than one receptor [1, 2]. CXCL8 belongs to a subgroup of chemokines known as ELR^+ chemokines because of the Glu4-Leu5-Arg6 amino acid sequence between positions 4 and 6. Other members of this group include CXCL1, 2, 3, 5, 6, and 7. A diverse variety of biological effects are attributed to CXCL8 and related ELR^+ chemokines, including several involving inflammatory cell activation and chemotaxis, production of reactive oxygen species, increased expression of the integrin CD11b–CD18, enhancement of cell adhesion to endothelial cells, promotion of angiogenesis, modulation of histamine and lipid mediator release as well as azurophil granule release [3]. The ELR^+ chemokines induce their effects through CXCR1 and CXCR2 expressed on a number of inflammatory cells in addition to human neutrophils, for example, T cells ($CD8^+$), B cells, basophils, dendritic cells, mast cells, activated monocytes, IL-2-activated natural killer cells and granulocyte-macrophage colony-stimulating factor (GM-CSF) and IL-3-stimulated eosinophils [4, 5]. Only CXCL8 and CXCL6 activate CXCR1 while the other ELR chemokines activate CXCR2 including CXCL1, 2, 3, 5, 6, 7, and 8.

Chemokine Biology – Basic Research and Clinical Application, Volume II
edited by Kuldeep Neote, Gordon L. Letts and Bernhard Moser

These chemokines are produced by many cells, including neutrophils [6, 7], monocytes [8], lymphocytes, macrophages, mast cells [9], vascular endothelial cells [10, 11], lung fibroblasts [11], keratinocytes [12], hepatocytes [13], stromal cells [14] and epithelial cells [15, 16] and this synthesis is stimulated by a wide variety of cytokines and other factors including IL-1β, TNF-α, lipopolysaccharide (LPS) and viral infection of cells [16, 17]. Recent studies have shown that metalloproteases (MMP9, neutrophil collagenase) which are released from activated neutrophils degrade extracellular matrix components and basement membranes thereby effecting cell migration. In addition, MMP9 can regulate chemokine activity by proteolytic processing [18]. MMP9 processes CXCL8 by N-terminal truncation, leaving ELR sequence intact, producing a protein with 10-fold higher potency than CXCL8 for its receptors resulting in a positive feedback [19]. This processing by MMP9 occurs with other ELR$^+$ chemokines including CXCL6 resulting in no change in activity and CXCL5 where there is first an increase in potency but further processing results in a decrease in potency for the receptors [19].

There is an evolving large body of evidence, some described below, indicating a role for ELR$^+$ chemokines acting through CXCR1 and CXCR2 that are directly or indirectly involved in the recruitment of many inflammatory cells to the site of inflammation [20].

The chemokines are present at elevated levels in several human inflammatory diseases and have been associated with pathophysiology; therefore agents that inhibit CXCR1 and/or CXCR2 will be useful to elucidate the role(s) of these receptors in inflammation and may demonstrate therapeutic benefit in some diseases.

Utility of CXCL8 antagonists in basic research

Identification of CXCL8 antagonists

First antagonists prepared, were modifications of the ligand by truncating the amino terminus with a couple amino acid substitutions. The resulting peptides had affinities close to that of native CXCL8 in binding studies and inhibited CXCL8-induced neutrophil elastase release and chemotaxis. They were dual antagonists able to inhibit CXCL1-, CXCL7- and CXCL8-induced responses [21]. More potent dual antagonists have been reported which are thought to be broad spectrum antagonists of ELR$^+$-CXC chemokines [22]. Continuation of this approach lead to selective CXCR2 antagonists which were able to selectively block CXCL8 binding to CXCR2 transfectants but not CXCR1. Functionally they blocked CXCL8-induced calcium mobilization only in CXCR2 transfectants [23]. There was limited *in vivo* data for these peptide antagonists, but inhibition in a dermal inflammation model was reported in the patent. Rodents which are shown to express only a functional CXCR2 receptor and use CINC as the ligand were administered a truncated antagonist of CINC and showed activity in a peritoneal zymosan-induced neutrophil assay [24].

Blocking mAbs to CXCR1 and CXCR2 have been used to demonstrate differential activities for the receptors. The neutrophil respiratory release was associated with CXCR1, while elastase release, calcium mobilization and chemotaxis were mediated through both receptors [23]. Several preclinical studies utilizing a fully humanized CXCL8 mAb supported the clinical evaluation of the mAb as an anti-inflammatory therapeutic [25, 26].

More recently with the use of high-throughput screening (HTS) by the pharmaceutical industry, small molecule antagonists have been identified for the CXCL8 receptors. The first reported antagonist using the HTS approach was by White et al. [27] with the identification of a selective CXCR2 antagonist. This compound (SB 225002) was effective at blocking CXCL8-induced neutrophil margination in rabbits. Another compound in this class, SB 272844, was reported to inhibit CXCL1- and CXCL8-induced apoptosis in human neutrophils [28]. Since then several other compounds from this series have been reported, all are CXCR2 selective, and have been shown to have activity in several animal models including LPS-induced lung neutrophilia [29], acute and chronic models of arthritis [30], wound healing [31] and lung inflammation [32]. The chemistry of this series has been extensively reported on [33, 34].

Several chemical series of CXCL8 antagonists have published in the literature, including Nicotinamide N-Oxides [35], Nicotinanilides [36], Thiazolethiols [37] and Amino-heteroaryl-quinoxalines [38]. Additionally, non-competitive, allosteric antagonists of CXCR1 and CXCR2 have been described [39]. One such compound, repertaxin, demonstrated activity in reperfusion injury models [40] (Tab. 1).

Preclinical models used for CXCR1 and CXCR2 antagonism

There are numerous preclinical models which have been used to elucidate a role of CXCR2 antagonists. However, most of these involve rodents which lack a functional CXCR1 receptor and a direct equivalent of CXCL8. This discrepancy makes it difficult to utilize rodent animal models with any degree of confidence that the CXCR2 antagonist activity observed will translate into humans. Nevertheless, many groups have used this species in an attempt to discover the role of CXCR2 in disease. Other species, such as the rabbit, represent a better alternative to the rodent as rabbits possess and express CXCR1 and CXCR2 and in addition express a CXCL8 homolog. Sequence homology of the CXCR1 and CXCR2 is high between species and some compounds have been shown to cross react with the CXCR2 derived from different species including rabbit, dog and ferret. Thus, the selection of a suitable animal model should represent the human system as close a possible in terms of ligand production and receptor expression.

Table 1 - Selected CXCR1 and or CXCR2 antagonist in development

Prototypic structure	Commentry
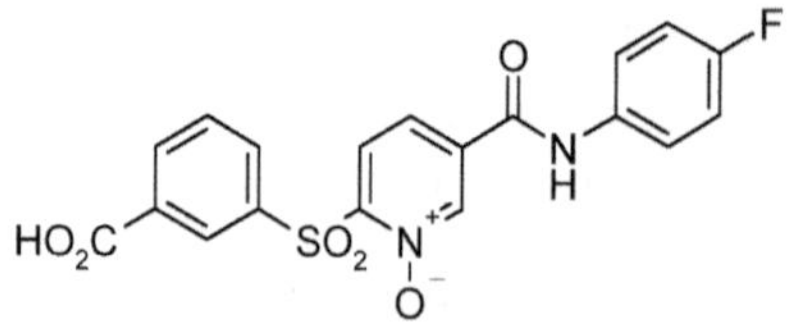	Celltech nicotinamide N-oxide derivative is a dual CXCR1/2 and FMLP antagonist active in the uM range.
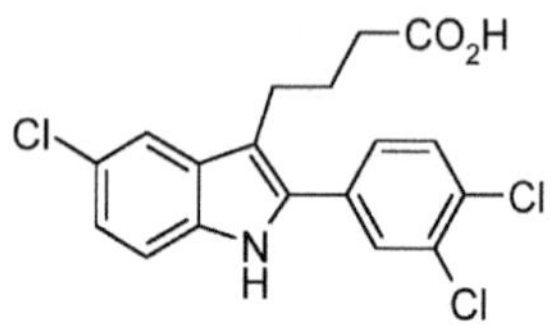	Fournier's non-peptide, orally-active, potent and specific IL-8 receptor antagonist. This series is a dual CXCR1, CXCR2 antagonist.
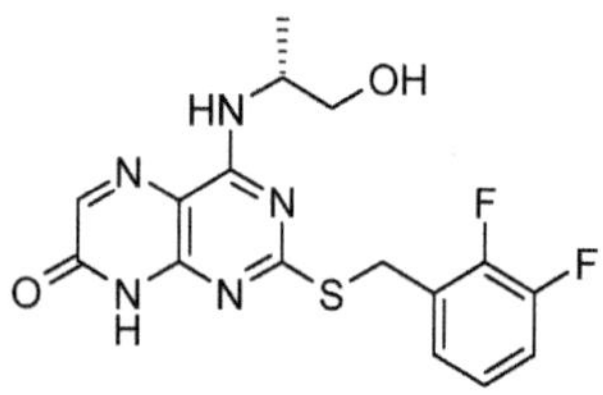	AstraZeneca has recently claimed two fused pyrimidine series-based antagonists selective for CXCR2. AZ has reportedly had 2 chemokine receptor antagonists in Phase I for COPD and RA.
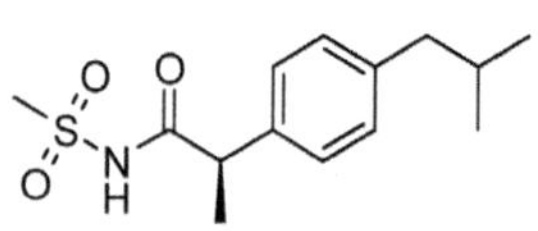	Dompe Repertaxin L-lysine salt is a non-competitive allosteric antagonist selective for CXCR1 and CXCR2. The compound is not orally bioavailable and is being contemplated for post-ischemic reperfusion injury.

Potential role and utility of CXCR2 antagonists in disease

Utility of CXCR2 antagonists in treating cancer

Many studies have implicated CXCL8, CXCL1 and the receptors CXCR1 and CXCR2 in angiogenesis and tumor growth. There are recent reviews published on the regulation of angiogenesis and metastases by the CXC chemokines [41, 42]. The ELR^{+} CXC chemokines that are angiogenic include CXCL1, CXCL2, CXCL3, CXCL5, CXCL6, CXCL7 and CXCL8 suggesting that CXCR2 is involved because only CXCL8 and CXCL6 activate CXCR1. An increased expression of these chemokines appears to correlate with increases in angiogenesis in a number of tumors (see review [43]) and there is increasing evidence that CXCL8 could play a role in malignant melanoma. Patients with metastatic melanoma have high concentrations of CXCL8 in their serum [44] and others have shown production of CXCL1 from melanoma cells lines. This factor appears to act in an autocrine growth factor [45]. CXCL8 has been shown to regulate angiogenesis while enhancing cell survival and proliferation [46]. Angiogenesis involves degradation of extracellular matrix and the matrix metalloproteinases have been shown to be elevated in tumors. CXCL8 has been shown to increase MMP2 in human melanoma cells and increases MMP2 and MMP9 expression and protein in human endothelial cells [47].

Mutations in the Ras oncogen are common in cancers and can lead to the overstimulation of NF-κB and AP-1 pathways. In a recent paper [48] Ras overexpression has been shown to stimulate the production of IL-8 from tumor cells. IL-8 then leads to the recruitment of inflammatory cells such as granulocytes and possibly endothelial cells and the establishment of an inflammatory response and neovascularization. Thus, inhibition of this cycle with an anti-IL-8 mAb or dual receptor antagonist would inhibit cell recruitment and vascularization.

CXCR2 appears to be the receptor involved in angiogenesis as it was shown to be receptor responsible for endothelial cell chemotaxis [49]. Studies with human vascular endothelial cells and mAbs to CXCR2 implicate this receptor in angiogenesis [50] and a recent study with mice deficient in CXCR2, demonstrated the role of this receptor in tumor growth and angiogenesis in lung cancer [51]. There are a significant number of studies suggesting a role for CXCR2 in tumor growth and this may be a direction for a future therapeutic opportunity.

Utility of CXCR2 antagonists in treating lung inflammation

COPD

In several lung disorders, including acute respiratory distress syndrome (ARDS), chronic obstructive pulmonary disease (COPD) and pulmonary fibrosis, there appears to be a significant inflammatory component of the disease. COPD is described as a fixed airways obstruction disease with a progressive decline in lung function. The primary etiological factor for the disease is cigarette smoking. COPD is characterized by a chronic inflammatory process in the lung where the main cells noted in the lung of patients are neutrophils, eosinophils, monocytes, macrophages and $CD8^+$ T cells. These cells express a number of chemokine receptors including CXCR1, CXCR2 and CCR2. Interestingly, monocytes express CXCR1 and $CD8^+$ T cells express CXCR1 and CXCR2 [52]. CXCL8 is a potent chemotactic factor in COPD as the chemotactic activity in patient bronchoalveolar lavage (BAL) fluid was strongly inhibited with a mAb for CXCL8 [53]. Inhibition of the monocyte recruitment may result in decreased macrophage presence in the COPD lung. Analysis of sputum and BAL samples, show an increase in the expression of CXCL8 [52, 54], CXCL1 [55] and CXCL5 [56] and the monocyte chemotactic factor CCL2 [55]. Furthermore, there is a strong correlation between the increased numbers of lung neutrophils and the chemokines CXCL1, CXCL5 and CXCL8. It has been suggested that these chemokines attract and activate neutrophils in the lung and promote the release of neutrophil-elastase and matrix metalloprotinases which in turn lead to lung fibrosis, airway stenosis and degradation of lung parenchyma [57]. CXCL1 is present at elevated levels in sputum of COPD patients and this ligand was shown to be a strong chemoattractant of monocytes as well as neutrophils.

Patients with COPD have exacerbations often associated with human rhinovirus infection. Levels of CXCL8 are significantly elevated in sputum during acute exacerbations and slowly decline after the exacerbations [58]. Recent data indicates that this CXCL8 increase with exacerbations is present in the large airways. The authors show that there is increased oxidative stress in airways of patients that increases further with exacerbations which is associated with increased CXCL8 and neutrophil numbers [59]. Interestingly, patients that exacerbate more often have higher CXCL8 sputum levels when the disease is stable, this may be a representation of the state of inflammation in the patients [60].

In addition, cigarette smoke leads to the inactivation of alpha-1 anti-trypsin which is a protective protease inhibitor in the lung [61]. Patients with alpha-1 anti-trypsin deficiency showed an elevated level of CXCL8 in BAL fluid early in the disease development. Analysis of forced expiratory breath condensate in COPD patients shows a strong negative correlation between lung ELR$^+$ chemokine levels and FEV_1. The beneficial effect of a small molecular weight compound in a rodent cigarette smoke-induced lung inflammation model was recently reported [26]; but to date there is no clinical data reported for any antagonist. However, it is proposed that such treatment would have a significant effect on reducing inflammatory cell recruitment and slow COPD progression [62].

ARDS

ARDS is an acute life threatening pulmonary disease with rapid onset of respiratory failure; this usually occurs following sepsis or a major clinical event. ARDS is characterized by an influx of protein rich edematous fluid into the lung as a result of increased permeability of the capillary-alveolar barrier. A role for chemokines in the pathophysiology of ARDS has been reviewed [1, 63, 64]. The major inflammatory cell infiltrating the lung in ARDS is the neutrophil. A recent review summarizes human and animal studies that help elucidate the neutrophil recruitment mechanisms involved in pulmonary vasculature [63]. Migration into the tissue is thought to be driven by ELR$^+$ chemokines, including CXCL8. Alveolar macrophages in the lung air space are probably responsible for the production of the chemokines including CXCL1, 2, 3, 5 and 8. The levels of the ELR^+ chemokines are elevated in patients at risk for or presently diagnosed with ARDS [64, 65]. In BAL fluid from ARDS patients the neutrophil chemotactic activity was inhibited by a monoclonal antibody to CXCL8. In addition, an antibody to CXCL1 was able to block about 50% of the chemotactic activity [65]. At times after the onset of ARDS CXCL5 is present in BAL fluid at concentrations greater than CXCL8 and is likely produced by airway epithelial cells.

A recent study indicates that major blunt trauma enhances neutrophil migratory activity possibly resulting from CXCL8 [66]. This suggests that the increased neutrophil migratory activity and elevated BAL fluid levels of CXCL8 may be critical for establishing the neutrophils in the lung, a characteristic of ARDS [66]. ARDS and acute lung injury (ALD) patients have elevated levels of α-macroglobulins and CXCL8 is present free and bound to α-macroglobulins [67]. This group was able to demonstrate that CXCL8 bound to α-macroglobulins retains its biological activity. Since the complexed CXCL8 may be protected from proteolysis, this form may be responsible for lung inflammation [67].

Airways inflammation in premature infants

Considerable interest has been focused on the role of neutrophils in lung damage in premature infants and the consequences of hyper-oxygen therapy. Inflammatory cell influx has been shown to precede the development of bronchopulmonary dysplasia and inflammation can exacerbate oxygen-induced lung injury [68]. Neutrophils contribute to the super oxide and protease induced stress, particularly in lung tissues. In a neonate rat model of hypoxia, antibodies to either CINC or MIP-2 significantly reduced the influx of neutrophils [69] and in addition reduced lung damage (septal thickening) but not elastase deposition [69]. Use of the small molecular weight CXCR2 selective antagonist SB-265610 (1–3 mg/kg) significantly reduced neutrophil infiltration and myeloperoxidase accumulation [70]. *In vitro* experiments showed that CINC-1 could protect rat neutrophils from time dependent apoptosis and that SB-265610 could reverse that protection [70]. This indicates that chemokines not only chemotax cells to the site of inflammation, but in addition, can prolong their survival by preventing apoptosis [71]. The inhibition of CXCL1- and CXCL8-induced apoptosis of human neutrophils with a selective antagonist was also demonstrated *in vitro* [28]. This highlights a new function for CXCR2 and selective antagonists which can reverse the anti-apoptotic effects of ELR^+ chemokines. Similar anti-apoptotic effects could take place in human lung cancer cell which have been reported to express both CXCR1 and 2 [72].

Utility of CXCR2 antagonists in psoriasis

Psoriasis is an inflammatory skin disorder most notably demarcated by an influx of T cells and neutrophils and the over production of several chemokines including CXCL1, CXCL8, [73] and CXCL9 [74]. High levels of CXCL1 and CXCL8 message were found in the upper epidermis expressed by both neutrophils and keratinocytes. In addition, CXCR1 was also found to be produced by papillary dermal cells associated with vessels [73]. CXCR5 was found to be absent in the psoriatic plaque. The overexpression of CXCL1 and CXCL8 could at least in part be responsible for the direct recruitment of neutrophils and T cells to this site. Discerning the role and importance of the various chemokines and their respective receptors in psoriasis or other dermatological conditions is difficult. Evidence using a fully humanized anti-IL-8 antibody (ABX-IL8) showed disappointing results which may point to the importance of the other CXCR1/2 chemokines found in psoriatic plaques. Thus a CXCR2 or dual receptor antibody may be required for efficacy in psoriasis. Additional data comes from *in vitro* experiments using human keratinocytes and the use of a standard medical treatment for the condition, most notably the exposure of these cells to UV radiation. In these experiments there was a preferential decrease in the expression of CXCR2 but not CXCR1 at 24 h post irradiation. Other supporting data for the role of chemokines in psoriasis comes from the use of infliximab (Anti-TNF-α) which leads to a reduction in symptoms and a decrease in the expression of chemokines most notably CXCL1, CXCL8 and CCL2 [75].

Utility of CXCR2 antagonists in rheumatoid arthritis

Rheumatoid arthritis (RA) is an inflammatory autoimmune condition characterized by the infiltration of inflammatory cells into the synovial tissue. Recent advances in the use of anti-TNF antibodies and receptor antagonists have defined TNF-α as a key mediator of synovial inflammation. *In vitro* studies have shown that TNF-α can lead to the generation of a number of inflammatory chemokines and other cytokines. This suggests that TNF-α is a central mediator of RA. Close examination of the clinical data however, points to a more complex picture for the roles of these mediators as only about 30% of RA patients reach ACR (American College of Rheumatologists) scores in excess of 70% with most patients sub-optimally responding to anti-TNF treatment. Use of the selective CXCR2 antagonist SB-265610 in a rabbit model of RA, lead to a reduction in multiple inflammatory mediators, including TNF-α, PGE2, LTB4, LTC4 and IL-8 produced in the knee joint [30]. In addition, there was a significant reduction in neutrophils, monocytes and lymphocytes infiltrating into the synovial fluid induced by both LPS and direct injection of IL-8. This strongly suggests that the recruitment of cells is either dependent on direct IL-8 chemotaxis or that one cell type, most probably neutrophils, conditions the synovial joint for the influx of other non-CXCR1/2 expressing cells. It is interesting to note that inhibition of cell infiltrates leads to the reduction in a number of non-cytokine inflammatory mediators including members of the prostaglandin and leukotriene families. Thus CXCR2 inhibitors may have multiple functions due to their ability to arrest cell infiltration. This data suggests that in the RA condition, cytokines and chemokines may not function in a linear fashion, with the highest cytokine (TNF-α) initiating the inflammatory response, and 'lower' cytokines being produced in response. It does however suggest that cytokines and chemokines interact with each other in a more circular fashion with cytokines initiating the inflammatory response and chemokines attracting the cells to the site. These two functions are dependent on each other and one factor cannot substitute for the other. Thus breaking this interdependence can lead to a reduction in cell trafficking and subsequently the production of cytokines, leukotrienes and other inflammatory mediators.

References

1 Puneet P, Moochhala S, Bhatia M (2005) Chemokines in acute respiratory distress syndrome. *Am J Physiol Lung Cell Mol Physiol* 288: L3–L15

2 Carter PH (2002) Chemokine receptor antagonism as an approach to anti-inflammatory therapy: 'just right' or plain wrong? *Curr Opin Chem Biol* 6: 510–525

3 Baggiolini M (1993) Chemotactic and inflammatory cytokines – CXC and CC proteins. *Adv Exp Med Biol* 351: 1–11

4 Baggiolini M, Dewald B, Moser B (1994) Interleukin-8 and related chemotactic cytokines – CXC and CC chemokines. *Adv Immunol* 55: 97–179

5 Mukaida N, Harada A, Matsushima K (1995) A novel leukocyte chemotactic and activating cytokine, interleukin-8 (IL-8). *Cancer Treat Res* 80: 261–286

6 Altstaedt J, Kirchner H, Rink L (1996) Cytokine production of neutrophils is limited to interleukin-8. *Immunology* 89: 563–568

7 Ryu J-S, Kang J-H, Jung S-Y, Shin M-H, Kim J-M, Park H, Min D-Y (2004) Production of interleukin-8 by human neutrophils stimulated with *Trichomonas vaginalis*. *Infect Immun* 72: 1326–1332

8 Arbabi S, Rosengart MR, Garcia I, Jelacic S, Maier RV (1999) Priming interleukin 8 production: role of platelet-activating factor and p38. *Arch Surg* 134: 1348–1353

9 Moller A, Lippert U, Lessmann D, Kolde G, Hamann K, Welker P, Schadendorf D, Rosenbach T, Luger T, Czarnetzki BM (1993) Human mast cells produce IL-8. *J Immunol* 151: 3261–3266

10 Burns MJ, Sellati TJ, Teng EI, Furie MB (1997) Production of interleukin-8 (IL-8) by cultured endothelial cells in response to *Borrelia burgdorferi* occurs independently of secreted [corrected] IL-1 and tumor necrosis factor alpha and is required for subsequent

transendothelial migration of neutrophils [published erratum appears in *Infect Immun* (1997) Jun; 65(6): 2508]. *Infect Immun* 65: 1217–1222

11 Smith RS, Fedyk ER, Springer TA, Mukaida N, Iglewski BH, Phipps RP (2001) IL-8 production in human lung fibroblasts and epithelial cells activated by the *Pseudomonas* autoinducer N-3-Oxododecanoyl homoserine lactone is transcriptionally regulated by NF-{{kappa}}B and Activator Protein-2. *J Immunol* 167: 366–374

12 Li J, Ireland GW, Farthing PM, Thornhill MH (1996) Epidermal and oral keratinocytes are induced to produce RANTES and IL-8 by cytokine stimulation. *J Invest Dermatol* 106: 661–666

13 Gomez-Quiroz LE, Paris R, Lluis JM, Bucio L, Souza V, Hernandez E, Gutierrez M, Santiago M, Garcia-Ruiz C, Fernandez-Checa JC et al (2005) Differential modulation of interleukin 8 by interleukin 4 and interleukin 10 in HepG2 cells treated with acetaldehyde. *Liver Int* 25: 122–130

14 Rougier F, Cornu E, Praloran V, Denizot Y (1998) IL-6 and IL-8 production by human bone marrow stromal cells. *Cytokine* 10: 93–97

15 Schulte R, Grassl GA, Preger S, Fessele S, Jacobi CA, Schaller M, Nelson PJ, Autenrieth IB (2000) *Yersinia enterocolitica* invasin protein triggers IL-8 production in epithelial cells via activation of Rel p65-p65 homodimers. *FASEB J* 14: 1471–1484

16 Subauste MC, Jacoby DB, Richards SM, Proud D (1995) Infection of a human respiratory epithelial cell line with rhinovirus. Induction of cytokine release and modulation of susceptibility to infection by cytokine exposure. *J Clin Invest* 96: 549–557

17 Kaplanski G, Teysseire N, Farnarier C, Kaplanski S, Lissitzky JC, Durand JM, Soubeyrand J, Dinarello CA, Bongrand P (1995) IL-6 and IL-8 production from cultured human endothelial cells stimulated by infection with *Rickettsia conorii* via a cell-associated IL-1 alpha-dependent pathway. *J Clin Invest* 96: 2839–2844

18 Van den Steen PE, Proost P, Wuyts A, Van Damme J, Opdenakker G (2000) Neutrophil gelatinase B potentiates interleukin-8 tenfold by aminoterminal processing, whereas it degrades CTAP-III, PF-4, and GRO-alpha and leaves RANTES and MCP-2 intact. *Blood* 96: 2673–2681

19 Van Den Steen PE, Wuyts A, Husson SJ, Proost P, Van Damme J, Opdenakker G (2003) Gelatinase B/MMP-9 and neutrophil collagenase/MMP-8 process the chemokines human GCP-2/CXCL6, ENA-78/CXCL5 and mouse GCP-2/LIX and modulate their physiological activities. *Eur J Biochem* 270: 3739–3749

20 Mukaida N (2000) Interleukin-8: an expanding universe beyond neutrophil chemotaxis and activation. *Int J Hematol* 72: 391–398

21 Moser B, Dewald B, Barella L, Schumacher C, Baggiolini M, Clark-Lewis I (1993) Interleukin-8 antagonists generated by N-terminal modification. *J Biol Chem* 268: 7125–7128

22 Li F, Zhang X, Gordon JR (2002) CXCL8((3-73))K11R/G31P antagonizes ligand binding to the neutrophil CXCR1 and CXCR2 receptors and cellular responses to CXCL8/IL-8. *Biochem Biophys Res Commun* 293: 939–944

23 Jones SA, Dewald B, Clark-Lewis I, Baggiolini M (1997) Chemokine antagonists that

discriminate between interleukin-8 receptors. Selective blockers of CXCR2. *J Biol Chem* 272: 16166–16169

24 Zagorski J, Wahl SM (1997) Inhibition of acute peritoneal inflammation in rats by a cytokine-induced neutrophil chemoattractant receptor antagonist. *J Immunol* 159: 1059–1062

25 Huang S, Mills L, Mian B, Tellez C, McCarty M, Yang XD, Gudas JM, Bar-Eli M (2002) Fully humanized neutralizing antibodies to interleukin-8 (ABX-IL8) inhibit angiogenesis, tumor growth, and metastasis of human melanoma. *Am J Pathol* 161: 125–134

26 Yang XD, Corvalan JR, Wang P, Roy CM, Davis CG (1999) Fully human anti-interleukin-8 monoclonal antibodies: potential therapeutics for the treatment of inflammatory disease states. *J Leukoc Biol* 66: 401–410

27 White JR, Lee JM, Young PR, Hertzberg RP, Jurewicz AJ, Chaikin MA, Widdowson J, Foley JJ, Martin LD, Griswold DE (1998) Identification of a potent, selective non-peptide CXCR2 antagonist that inhibits interleukin-8-induced neutrophil migration. *J Biol Chem* 273: 10095–10098

28 Glynn PC, Henney E, Hall IP (2002) The selective CXCR2 antagonist SB272844 blocks interleukin-8 and growth-related oncogene-alpha-mediated inhibition of spontaneous neutrophil apoptosis. *Pulm Pharmacol Ther* 15: 103–110

29 Hay DW, Sarau HM (2001) Interleukin-8 receptor antagonists in pulmonary diseases. *Curr Opin Pharmacol* 1: 242–247

30 Podolin PL, Bolognese BJ, Foley JJ, Schmidt DB, Buckley PT, Widdowson KL, Jin Q, White JR, Lee JM, Goodman RB et al (2002) A potent and selective nonpeptide antagonist of CXCR2 inhibits acute and chronic models of arthritis in the rabbit. *J Immunol* 169: 6435–6444

31 Milatovic S, Nanney LB, Yu Y, White JR, Richmond A (2003) Impaired healing of nitrogen mustard wounds in CXCR2 null mice. *Wound Repair Regen* 11: 213–219

32 Stevenson CS, Coote K, Webster R, Johnston H, Atherton HC, Nicholls A, Giddings J, Sugar R, Jackson A, Press NJ et al (2005) Characterization of cigarette smoke-induced inflammatory and mucus hypersecretory changes in rat lung and the role of CXCR2 ligands in mediating this effect. *Am J Physiol Lung Cell Mol Physiol* 288: L514–L522

33 Jin Q, Nie H, McCleland BW, Widdowson KL, Palovich MR, Elliott JD, Goodman RM, Burman M, Sarau HM, Ward KW et al (2004) Discovery of potent and orally bioavailable N,N'-diarylurea antagonists for the CXCR2 chemokine receptor. *Bioorg Med Chem Lett* 14: 4375–4378

34 Widdowson KL, Elliott JD, Veber DF, Nie H, Rutledge MC, McCleland BW, Xiang JN, Jurewicz AN, Hertzberg RP, Foley JJ et al (2004) Evaluation of potent and selective small-molecule antagonists for the CXCR2 chemokine receptor. *J Med Chem* 47: 1319–1321

35 Cutshall NS, Ursino R, Kucera KA, Latham J, Ihle NC (2001) Nicotinamide N-oxides as CXCR2 antagonists. *Bioorg Med Chem Lett* 11: 1951–1954

36 Cutshall NS, Kucera KA, Ursino R, Latham J, Ihle NC (2002) Nicotinanilides as inhibitors of neutrophil chemotaxis. *Bioorg Med Chem Lett* 12: 1517–1520

37 Baxter A, Bennion C, Bent J, Boden K, Brough S, Cooper A, Kinchin E, Kindon N, McInally T, Mortimore M et al (2003) Hit-to-lead studies: the discovery of potent, orally bioavailable triazolethiol CXCR2 receptor antagonists. *Bioorg Med Chem Lett* 13: 2625–2628

38 Li JJ, Carson KG, Trivedi BK, Yue WS, Ye Q, Glynn RA, Miller SR, Connor DT, Roth BD, Luly JR et al (2003) Synthesis and structure-activity relationship of 2-amino-3-heteroaryl-quinoxalines as non-peptide, small-molecule antagonists for interleukin-8 receptor. *Bioorg Med Chem* 11: 3777–3790

39 Casilli F, Bianchini A, Gloaguen I, Biordi L, Alesse E, Festuccia C, Cavalieri B, Strippoli R, Cervellera MN, Di Bitondo R et al (2005) Inhibition of interleukin-8 (CXCL8/IL-8) responses by repertaxin, a new inhibitor of the chemokine receptors CXCR1 and CXCR2. *Biochem Pharmacol* 69: 385–394

40 Bertini R, Allegretti M, Bizzarri C, Moriconi A, Locati M, Zampella G, Cervellera MN, Di Cioccio V, Cesta MC, Galliera E et al (2004) Noncompetitive allosteric inhibitors of the inflammatory chemokine receptors CXCR1 and CXCR2: prevention of reperfusion injury. *Proc Natl Acad Sci USA* 101: 11791–11796

41 Strieter RM, Belperio JA, Burdick MD, Sharma S, Dubinett SM, Keane MP (2004) CXC chemokines: angiogenesis, immunoangiostasis, and metastases in lung cancer. *Ann NY Acad Sci* 1028: 351–360

42 Strieter RM, Belperio JA, Phillips RJ, Keane MP (2004) CXC chemokines in angiogenesis of cancer. *Semin Cancer Biol* 14: 195–200

43 Rosenkilde MM, Schwartz TW (2004) The chemokine system – a major regulator of angiogenesis in health and disease. *Apmis* 112: 481–495

44 Scheibenbogen C, Mohler T, Haefele J, Hunstein W, Keilholz U (1995) Serum interleukin-8 (IL-8) is elevated in patients with metastatic melanoma and correlates with tumour load. *Melanoma Res* 5: 179–181

45 Haghnegahdar H, Du J, Wang D, Strieter RM, Burdick MD, Nanney LB, Cardwell N, Luan J, Shattuck-Brandt R, Richmond A (2000) The tumorigenic and angiogenic effects of MGSA/GRO proteins in melanoma. *J Leukoc Biol* 67: 53–62

46 Li A, Dubey S, Varney ML, Dave BJ, Singh RK (2003) IL-8 directly enhanced endothelial cell survival, proliferation, and matrix metalloproteinases production and regulated angiogenesis. *J Immunol* 170: 3369–3376

47 McCawley LJ, Matrisian LM (2000) Matrix metalloproteinases: multifunctional contributors to tumor progression. *Mol Med Today* 6: 149–156

48 Sparmann A, Bar-Sagi D (2004) Ras-induced interleukin-8 expression plays a critical role in tumor growth and angiogenesis. *Cancer Cell* 6: 447–458

49 Addison CL, Daniel TO, Burdick MD, Liu H, Ehlert JE, Xue YY, Buechi L, Walz A, Richmond A, Strieter RM (2000) The CXC chemokine receptor 2, CXCR2, is the putative receptor for ELR^+ CXC chemokine-induced angiogenic activity. *J Immunol* 165: 5269–5277

50 Heidemann J, Ogawa H, Dwinell MB, Rafiee P, Maaser C, Gockel HR, Otterson MF, Ota DM, Lugering N, Domschke W et al (2003) Angiogenic effects of interleukin 8

(CXCL8) in human intestinal microvascular endothelial cells are mediated by CXCR2. *J Biol Chem* 278: 8508–8515
51 Keane MP, Belperio JA, Xue YY, Burdick MD, Strieter RM (2004) Depletion of CXCR2 inhibits tumor growth and angiogenesis in a murine model of lung cancer. *J Immunol* 172: 2853–2860
52 Yamamoto C, Yoneda T, Yoshikawa M, Fu A, Tokuyama T, Tsukaguchi K, Narita N (1997) Airway inflammation in COPD assessed by sputum levels of interleukin-8. *Chest* 112: 505–510
53 Hill AT, Bayley D, Stockley RA (1999) The interrelationship of sputum inflammatory markers in patients with chronic bronchitis. *Am J Respir Crit Care Med* 160: 893–898
54 Keatings VM, Collins PD, Scott DM, Barnes PJ (1996) Differences in interleukin-8 and tumor necrosis factor-alpha in induced sputum from patients with chronic obstructive pulmonary disease or asthma. *Am J Respir Crit Care Med* 153: 530–534
55 Traves SL, Culpitt SV, Russell RE, Barnes PJ, Donnelly LE (2002) Increased levels of the chemokines GROalpha and MCP-1 in sputum samples from patients with COPD. *Thorax* 57: 590–595
56 Pesci A, Balbi B, Majori M, Cacciani G, Bertacco S, Alciato P, Donner CF (1998) Inflammatory cells and mediators in bronchial lavage of patients with chronic obstructive pulmonary disease. *Eur Respir J* 12: 380–386
57 Williams TJ, Jose PJ (2001) Neutrophils in chronic obstructive pulmonary disease. *Novartis Found Symp* 234: 136–141
58 Aaron SD, Angel JB, Lunau M, Wright K, Fex C, Le Saux N, Dales RE (2001) Granulocyte inflammatory markers and airway infection during acute exacerbation of chronic obstructive pulmonary disease. *Am J Respir Crit Care Med* 163: 349–355
59 Drost EM, Skwarski KM, Sauleda J, Soler N, Roca J, Agusti A, MacNee W (2005) Oxidative stress and airway inflammation in severe exacerbations of COPD. *Thorax* 60: 293–300
60 Wedzicha JA (2001) Mechanisms of exacerbations. *Novartis Found Symp* 234: 84–93
61 Pryor WA, Dooley MM, Church DF (1985) Mechanisms of cigarette smoke toxicity: the inactivation of human alpha-1-proteinase inhibitor by nitric oxide/isoprene mixtures in air. *Chem Biol Interact* 54: 171–183
62 Barnes PJ (2003) Cytokine-directed therapies for the treatment of chronic airway diseases. *Cytokine Growth Factor Rev* 14: 511–522
63 Reutershan J, Ley K (2004) Bench-to-bedside review: acute respiratory distress syndrome – how neutrophils migrate into the lung. *Crit Care* 8: 453–461
64 Martin TR (1999) Lung cytokines and ARDS: Roger S. Mitchell Lecture. *Chest* 116: 2S–8S
65 Goodman RB, Strieter RM, Martin DP, Steinberg KP, Milberg JA, Maunder RJ, Kunkel SL, Walz A, Hudson LD, Martin TR (1996) Inflammatory cytokines in patients with persistence of the acute respiratory distress syndrome. *Am J Respir Crit Care Med* 154: 602–611
66 Pallister I, Dent C, Topley N (2002) Increased neutrophil migratory activity after major

trauma: a factor in the etiology of acute respiratory distress syndrome? *Crit Care Med* 30: 1717–1721

67 Kurdowska AK, Geiser TK, Alden SM, Dziadek BR, Noble JM, Nuckton TJ, Matthay MA (2002) Activity of pulmonary edema fluid interleukin-8 bound to alpha(2)-macroglobulin in patients with acute lung injury. *Am J Physiol Lung Cell Mol Physiol* 282: L1092–L1098

68 Jobe AH, Ikegami M (1998) Mechanisms initiating lung injury in the preterm. *Early Hum Dev* 53: 81–94

69 Deng H, Mason SN, Auten RL Jr (2000) Lung inflammation in hyperoxia can be prevented by antichemokine treatment in newborn rats. *Am J Respir Crit Care Med* 162: 2316–2323

70 Auten RL, Richardson RM, White JR, Mason SN, Vozzelli MA, Whorton MH (2001) Nonpeptide CXCR2 antagonist prevents neutrophil accumulation in hyperoxia-exposed newborn rats. *J Pharmacol Exp Ther* 299: 90–95

71 Broxmeyer HE, Kohli L, Kim CH, Lee Y, Mantel C, Cooper S, Hangoc G, Shaheen M, Li X, Clapp DW (2003) Stromal cell-derived factor-1/CXCL12 directly enhances survival/antiapoptosis of myeloid progenitor cells through CXCR4 and G(alpha)i proteins and enhances engraftment of competitive, repopulating stem cells. *J Leukoc Biol* 73: 630–638

72 Zhu YM, Webster SJ, Flower D, Woll PJ (2004) Interleukin-8/CXCL8 is a growth factor for human lung cancer cells. *Br J Cancer* 91: 1970–1976

73 Gillitzer R, Ritter U, Spandau U, Goebeler M, Brocker EB (1996) Differential expression of GRO-alpha and IL-8 mRNA in psoriasis: a model for neutrophil migration and accumulation *in vivo*. *J Invest Dermatol* 107: 778–782

74 Goebeler M, Toksoy A, Spandau U, Engelhardt E, Brocker EB, Gillitzer R (1998) The C-X-C chemokine Mig is highly expressed in the papillae of psoriatic lesions. *J Pathol* 184: 89–95

75 Benoit S, Toksoy A, Brocker EB, Gillitzer R, Goebeler M (2004) Treatment of recalcitrant pustular psoriasis with infliximab: effective reduction of chemokine expression. *Br J Dermatol* 150: 1009–1012

Current status of CCR1 antagonists in clinical trials

Ronald P. Gladue and Matthew F. Brown

Pfizer Global Research and Development, Department of Antibacterials, Immunology, and Cancer; MS 8220-2410, Eastern Point Road, Groton, CT 06340, USA

Introduction

The chemokine receptor, CCR1, is believed to play a crucial role in the migration of leukocytes to sites of inflammation. It has been shown to be expressed on monocytes, T cells, dendritic cells, and in some cases, neutrophils [1–4]) and interacts with at least 7 different ligands including CCL3 (MIP-1α, macrophage inflammatory protein-1α), CCL5 (RANTES, regulated on activation, normal T cell expressed and secreted), CCL7 (MCP-3, monocyte chemotactic protein-3), CCL14 (HCC-1, hemofiltrate C-C chemokine-1), CCL8 (MCP-2, monocyte chemotactic protein-2), CCL15 (leukotactin-1), and CCL23 (MPIF, myeloid progenitor inhibitory factor-1) [5, 6]. These ligands have been shown to have potent chemotactic activity *in vitro* [4] and, in some cases *in vivo* where intradermal injection of CCL3 or CCL5 induced a robust cellular infiltration [3, 7]. Further, these chemokines can be produced by the very cells they attract to inflammatory sites. For example, peripheral blood monocytes can secrete CCL3 following activation, potentially setting up an amplification loop whereby monocytes migrate into tissue in response to CCR1, then become activated and secrete CCR1 ligands such as CCL3, thereby recruiting more cells and setting up a state of chronic inflammation. These properties suggest that CCR1 may play an important role in perpetuating inflammatory responses.

In addition to mediating cell migration, CCR1 signaling has been shown to upregulate integrins such as Mac-1 (CD11b), thus promoting the firm adherence of leukocytes to the endothelium [8]. CCR1 signaling may also contribute to tissue damage and inflammation through the enhancement of T cell activation [9], regulation of TH-1/TH-2 polarization [10, 11] and stimulation of macrophage function [12] and protease secretion [8, 13, 14]. Taken together, these properties support CCR1 as an attractive therapeutic target to modulate leukocyte infiltration and decrease the associated tissue damage common to autoimmune diseases.

Chemokine Biology – Basic Research and Clinical Application, Volume II
edited by Kuldeep Neote, Gordon L. Letts and Bernhard Moser

Evidence for the role of CCR1 in rheumatoid arthritis

Rheumatoid arthritis (RA) is a chronic inflammatory disease affecting 0.5–2% of the population in the western world, the majority of whom are female. Central to the pathogenesis of this disease is the infiltration of monocytes into synovial tissue. This is supported by the predominance of monocytes found in the joint during flare [15, 16], the role of monocyte-derived proinflammatory mediators in disease progression (e.g., TNF, IL-1) [17, 18] and the ability of monocytes to secrete tissue damaging proteolytic enzymes that participate in joint destruction [19]. Two chemokines thought to play a major role in the recruitment of monocytes into synovial tissues are the CCR1 ligands CCL3 and CCL5. Evidence in support of this is provided by a number of studies which have demonstrated an elevation of these chemokines in the synovial tissue and fluid of RA patients [20–24]. In support of their role in the pathogenesis of disease, the level of CCL3 and monocytes in synovial tissue were shown to be directly proportional to the magnitude of joint pain [16]. The role of CCR1 in the pathogenesis of RA is also supported by human genetic association studies [25] and by animal models of arthritis [26–28].

Evidence for the role of CCR1 in multiple sclerosis

Multiple sclerosis (MS) is a chronic, progressive, immune-mediated disease of the central nervous system. The disease course is highly variable but characterized by initial demyelination of nerve fiber followed by axonal loss. Key leukocytes believed to be involved in the pathogenesis of this disease include autoreactive T cells, which may initiate the disease, and monocytes, which may be responsible for the demyelination [29]. As such, inhibition of T cell and monocyte infiltration into the central nervous system (CNS) may provide a viable new strategy for disease treatment. The first clinical evidence in support of this strategy was derived from Phase 2 studies conducted with natalizumab; an $\alpha4$ integrin antibody which blocks leukocyte infiltration into the CNS resulting in robust efficacy in MS patients [30].

Evidence for the role of CCR1 and its ligands in MS include their expression in demyelinating lesions, animal model data, and genetic association studies. Analysis of the cerebrospinal fluid of MS patients demonstrated increased levels of CCR1 during early and acute demyelinating stages as well as during relapse [31]. In addition, the CCR1 ligands, CCL3 and CCL5, have been detected in active regions of demyelination [32, 33] and CCL5 has been reported to be elevated in the cerebrospinal fluid of MS patients [34]. Additional evidence for the role of CCR1 in the pathogenesis of MS is provided by studies conducted in an animal model of CNS inflammation, experimental autoimmune encephalomyelitis (EAE). In this model, inhibition of the CCR1 ligands, CCL3 and CCL5 have been shown to decrease disease severity [35, 36]. Further, CCR1$^{-/-}$ animals have shown reduced levels of leukocyte infiltration into the CNS as well as decreased paralysis in this model [37].

Evidence for role of CCR1 in other diseases

In addition to MS and RA, CCR1 may also play a role in other diseases. CCR1 ligands have been shown to be upregulated in tissue transplants [38] and $CCR1^{-/-}$ mice experience delayed rejection of heart transplants *versus* wild-type animals [39]. In multiple myeloma, CCR1 ligands have been suggested to play a role in bone destruction, by controlling the migration of osteoclasts [40, 41]. Finally, recent reports demonstrate that CCR1 expression is elevated on neurites associated with Aβ42 positive plaques in the brain of Alzheimer's patients. Further, this expression seemed to correlate with the severity of disease suggesting its potential use as an early marker of Alzheimer's disease and raising questions as to the significance of this expression on disease pathogenesis [42]. Collectively, these findings suggest that a CCR1 antagonist may be useful to treat a number of diseases.

Small molecule antagonists of CCR1

A number of pharmaceutical companies have disclosed efforts to identify small molecule CCR1 antagonists, including Pfizer, Berlex Biosciences/Schering AG, Banyu, Millennium, Novartis and Chemocentryx/Forrest Labs to name a few (Tab. 1). The majority of the chemical series explored thus far are structurally related in that they incorporate a positively charged nitrogen; a feature common to many small molecule antagonists of G-protein coupled receptors. Hydroxy piperidines (example 1) and benzyl piperazines (BX471), as well as xanthene carboxamides (example 2) are good examples of series that incorporate this structural feature. Alternatively, Pfizer has disclosed a structurally novel series of antagonists discovered in the course of screening their compound library (CP-481,715). This series lacks the basic nitrogen, and, in fact is derived from a chemical series originally designed for a renin inhibitor program. While numerous companies have been involved in CCR1 antagonist drug discovery, to date, few have reported results from human clinical trials. The exceptions are Berlex Biosciences/Schering AG (BX-471 / ZK811792) [43] and Pfizer (CP-481,715) [44].

BX471 (ZK-811752)

Schering AG and Berlex Biosciences have advanced their lead candidate, BX471 (ZK-811752), into clinical trials for psoriasis and multiple sclerosis [43, 45]. This

Table 1 CCR1 antagonists. Shown are published CCR1 antagonists in preclinical and clinical development

Compound	Company
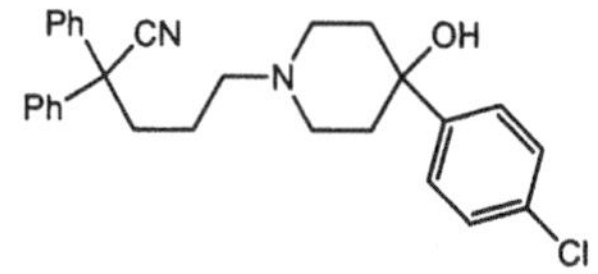 Example 1	Berlex [53]
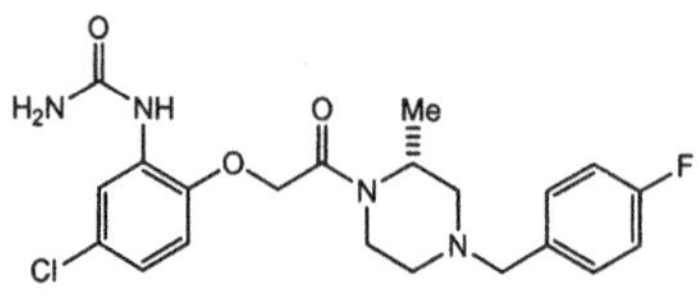 BX471	Berlex [45]
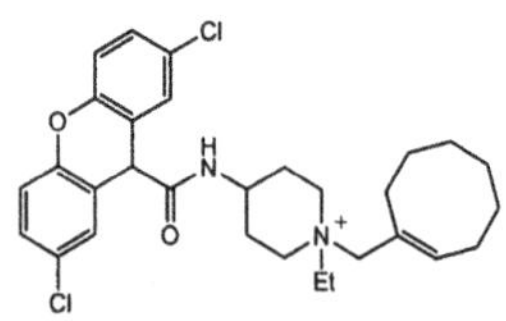 Example 2	Banyu [54]
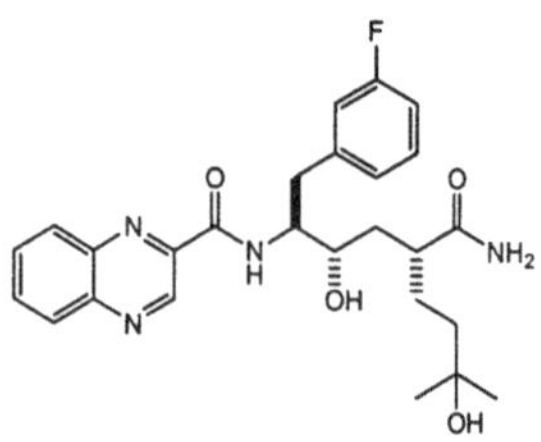 CP-481,715	Pfizer [8, 50]

compound is a potent and selective CCR1 antagonist that was shown to be effective in a number of animal disease models, including models of MS [45], transplantation [46, 47] and renal fibrosis [48]. BX471 has completed Phase I clinical trials and Phase II psoriasis studies, however, the results of these studies have not been published. Schering AG has also initiated Phase I/II clinical trials to determine if a radiolabeled CCR1 antagonist could be used as a biomarker to track disease progression in Alzheimer's patients [42].

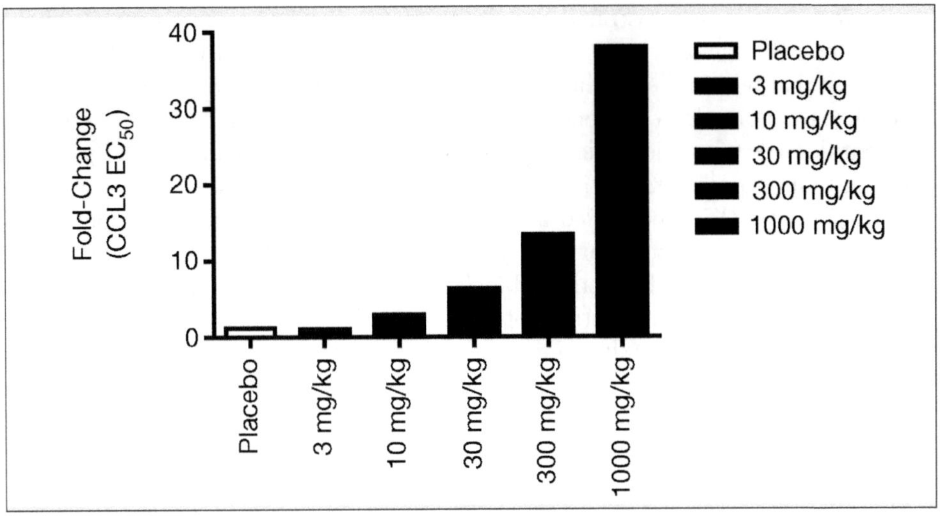

Figure 1
Inhibition of CCL3-induced CD11b upregulation on monocytes in blood taken from individuals dosed with CP-481,715. Individuals received one dose of either 3, 10, 30, 300 or 1000 mg CP-481,715. One hour later, peripheral blood was collected and incubated with various concentrations of CCL3. The EC_{50} of CCL3 to induce CD11b upregulation on monocytes was then determined. The data represents the average amount of CCL3 necessary to increase CD11b expression on monocytes for each dose group and is expressed as the fold-change in CCL3 EC_{50} over pre-dose levels.

CP-481,715

Pfizer, Inc. has described a structurally novel small molecule CCR1 antagonist, CP-481,715 [8, 49–51]. CP-481,715 was found to block the binding and function of all CCR1 ligands and retain good potency in human whole blood; an important attribute for a therapeutic agent. Further, CP-481,715 was selective for CCR1, when profiled against a panel of 41 chemokine and non-chemokine G-protein coupled receptors (GPCRs), and was also selective for the human receptor, showing no CCR1 antagonist activity on mouse, rat, or monkey cells.

Phase I studies with CP-481,715 have been completed. Whole blood taken from individuals dosed orally with CP-481,715 and stimulated with various concentrations of CCL3 demonstrated a dose related shift in the amount of CCL3 necessary to upregulate Mac-1 (CD11b) on monocytes (Fig. 1). This rightward shift in the dose response curve illustrated that CP-481,715 was active in the blood of dosed individuals, was able to inhibit CCL3 induced monocyte activation, and was acting as a competitive antagonist.

In vitro studies demonstrated the ability of CP-481,715 to inhibit 90% of the monocyte chemotactic activity induced by synovial fluid taken from RA patients, suggesting that within the milieu of chemokines present in synovial fluid, CCR1 ligands play a major role in monocyte migration [8]. This prompted a 2 week study in rheumatoid arthritis patients designed to assess whether blockade of CCR1 would decrease monocyte infiltration into synovial tissue [52]. In this study, patients treated with CP-481,715 experienced a significant reduction in synovial tissue monocytes *versus* placebo thus providing the first clinical evidence that a chemokine receptor antagonist could effectively block leukocyte infiltration into an inflammatory site. However, the implications of decreasing monocyte infiltration are unclear since a 6 week Phase II study conducted with CP-481,715 in RA patients did not demonstrate clinical efficacy, despite good toleration [44]. However, future studies would be needed to assess whether higher dose levels and/or a longer duration of CCR1 blockade is necessary.

Issues and challenges to the development of chemokine receptor antagonists

There are several challenges unique to the development of CCR1 antagonists. One particularly troublesome characteristic is a high degree of specificity for the human receptor [8, 45]. This creates two major issues. First, the lack of animal cross-activity makes it difficult to assess mechanism related toxicity in preclinical studies. This is a particular problem for the assessment of any potential effects on reproduction, an issue for advancing a compound into diseases which afflict women of child-bearing potential. By necessity, the lack of cross-activity impedes clinical development due to the need to conduct very cautious dose escalation studies. Second, the lack of animal cross-activity can prevent the use of standard animal disease models, which eliminates an important tool for target validation and also complicates efforts to predict efficacious human plasma concentrations.

Another challenge associated with the development of CCR1 antagonists is the ability of its ligands to interact with multiple receptors. For example, CCL3 and CCL5 can also interact with CCR5, whereas CCL7 and CCL8 can also interact with CCR2. This promiscuity of ligand interactions with other receptors could suggest that blockade of one receptor may be insufficient for efficacy, or that over time, alternative receptors for the ligands may compensate. The results of ongoing clinical trials with various chemokine receptor antagonists will help address this possibility.

The development of CCR1 antagonists is also complicated by the significant differences in the expression and function of CCR1 in animals as compared to humans. For example, in mice, it has been shown that CCR1 is an important chemotactic factor for neutrophils but not monocytes [10]. In contrast, in human, CCR1 is an important chemotactic factor for human monocytes [8, 52] and is generally not considered a potent chemotactic receptor for human neutrophils, although its ligand, CCL3, did promote neutrophil infiltration following intradermal injection [3]. These differences can lead to the misinterpretation of results from animal disease models, and can also limit our ability to determine which chemokine receptor(s) are the appropriate therapeutic targets for a given disease. As such, the role of chemokines and their receptors in human disease can only be definitively assessed in human clinical trials.

Conclusion

CCR1 antagonists represent a new class of agents that appear to be well tolerated, demonstrate efficacy in numerous animal models, and have the ability to selectively decrease cell infiltration in human disease. The results from ongoing clinical trials in rheumatoid arthritis, psoriasis, and multiple sclerosis patients will hopefully uncover the therapeutic potential of these novel agents.

References

1 Bonecchi R, Polentarutti N, Luini W, Borsatti A, Bernasconi S, Locati M, Power C, Proudfoot A, Wells TN, Mackay C et al (1999) Up-regulation of CCR1 and CCR3 and induction of chemotaxis to CC chemokines by IFN-gamma in human neutrophils. *J Immunol* 162: 474–479

2 Struyf S, Menten P, Lenaerts JP, Put W, D'Haese A, De Clercq E, Schols D, Proost P, Van Damme J (2001) Diverging binding capacities of natural LD78beta isoforms of macrophage inflammatory protein-1alpha to the CC chemokine receptors 1, 3 and 5 affect their anti-HIV-1 activity and chemotactic potencies for neutrophils and eosinophils. *Eur J Immunol* 31: 2170–2178

3 Lee SC, Brummet ME, Shahabuddin S, Woodworth TG, Georas SN, Leiferman KM, Gilman SC, Stellato C, Gladue RP, Schleimer RP et al (2000) Cutaneous injection of human subjects with macrophage inflammatory protein-1 alpha induces significant recruitment of neutrophils and monocytes. *J Immunol* 164: 3392–3401

4 Premack BA, Schall TJ (1996) Chemokine receptors: gateways to inflammation and infection. *Nat Med* 2: 1174–1178

5 Zlotnik A, Yoshie O (2000) Chemokines: a new classification system and their role in immunity. *Immunity* 12: 121–127

6 Tsou CL, Gladue RP, Carroll LA, Paradis T, Boyd JG, Nelson RT, Neote K, Charo IF (1998) Identification of C-C chemokine receptor 1 (CCR1) as the monocyte hemofiltrate C-C chemokine (HCC)-1 receptor. *J Exp Med* 188: 603–608

7 Beck LA, Dalke S, Leiferman KM, Bickel CA, Hamilton R, Rosen H, Bochner BS,

Schleimer RP (1997) Cutaneous injection of RANTES causes eosinophil recruitment: comparison of nonallergic and allergic human subjects. *J Immunol* 159: 2962–2972

8 Gladue RP, Tylaska LA, Brissette WH, Lira PD, Kath JC, Poss CS, Brown MF, Paradis TJ, Conklyn MJ, Ogborne KT et al (2003) CP-481,715, a potent and selective CCR1 antagonist with potential therapeutic implications for inflammatory diseases. *J Biol Chem* 278: 40473–40480

9 Ward SG, Bacon K, Westwick J (1998) Chemokines and T lymphocytes: More than an attraction. *Immunity* 9: 1–11

10 Gao JL, Wynn TA, Chang Y, Lee EJ, Broxmeyer HE, Cooper S, Tiffany HL, Westphal H, Kwon-Chung J, Murphy PM (1997) Impaired host defense, hematopoiesis, granulomatous inflammation and type 1-type 2 cytokine balance in mice lacking CC chemokine receptor 1. *J Exp Med* 185: 1959–1968

11 Colantonio L, Iellem A, Clissi B, Pardi R, Rogge L, Sinigaglia F, D'Ambrosio D (1999) Upregulation of integrin alpha6/beta1 and chemokine receptor CCR1 by interleukin-12 promotes the migration of human type 1 helper T cells. *Blood* 94: 2981–2989

12 Fahey TJ 3rd, Tracey KJ, Tekamp-Olson P, Cousens LS, Jones WG, Shires GT, Cerami A, Sherry B (1992) Macrophage inflammatory protein 1 modulates macrophage function. *J Immunol* 148: 2764–2769

13 Klier CM, Nelson EL, Cohen CD, Horuk R, Schlondorff D, Nelson PJ (2001) Chemokine-Induced secretion of gelatinase B in primary human monocytes. *Biol Chem* 382: 1405–1410

14 Robinson SC, Scott KA, Balkwill FR (2002) Chemokine stimulation of monocyte matrix metalloproteinase-9 requires endogenous TNF-alpha. *Eur J Immunol* 32: 404–412

15 Cutolo M, Sulli A, Barone A, Seriolo B, Accardo S (1993) Macrophages, synovial tissue and rheumatoid arthritis. *Clin Exp Rheum* 11: 331–339

16 Tak PP, Smeets TJ, Daha MR, Kluin PM, Meijers KA, Brand R, Meinders AE, Breedveld FC (1997) Analysis of the synovial cell infiltrate in early rheumatoid synovial tissue in relation to local disease activity. *Arth Rheum* 40: 217–225

17 Dayer JM (2002) The saga of the discovery of IL-1 and TNF and their specific inhibitors in the pathogenesis and treatment of rheumatoid arthritis. *Joint, Bone, Spine: Revue du Rhumatisme* 69: 123–132

18 Meyer O (2000) Place des therapeutiques anti-TNF dans la polyarthrite rhumatoide. *Presse Medicale* 29: 463–468

19 Jenkins JK, Hardy KJ (2002) Biological modifier therapy for the treatment of rheumatoid arthritis. *Am J Med Sci* 323: 197–205

20 Katrib A, Tak PP, Bertouch JV, Cuello C, McNeil HP, Smeets TJ, Kraan MC, Youssef PP (2001) Expression of chemokines and matrix metalloproteinases in early rheumatoid arthritis. *Rheumatology* 40: 988–994

21 Koch AE, Kunkel SL, Harlow LA, Mazarakis DD, Haines GK, Burdick MD, Pope RM, Strieter RM (1994) Macrophage inflammatory protein-1 alpha. A novel chemotactic cytokine for macrophages in rheumatoid arthritis. *J Clin Invest* 93: 921–928

22 Volin MV, Shah MR, Tokuhira M, Haines GK, Woods JM, Koch AE (1998) RANTES

expression and contribution to monocyte chemotaxis in arthritis. *Clin Immunol Immunopathol* 89: 44–53
23 Ellingsen T, Buus A, Moller BK, Stengaard-Pedersen K (2000) *In vitro* migration of mononuclear cells towards synovial fluid and plasma from rheumatoid arthritis patients correlates to RANTES synovial fluid levels and to clinical pain parameters. *Scand J Rheum* 29: 216–221
24 al-Mughales J, Blyth TH, Hunter JA, Wilkinson PC (1996) The chemoattractant activity of rheumatoid synovial fluid for human lymphocytes is due to multiple cytokines. *Clin Exp Immunol* 106: 230–236
25 Makki RF, al Sharif F, Gonzalez-Gay MA, Garcia-Porrua C, Ollier WE, Hajeer AH (2000) RANTES gene polymorphism in polymyalgia rheumatica, giant cell arteritis and rheumatoid arthritis. *Clin Exp Rheum* 18: 391–393
26 Plater-Zyberk C, Hoogewerf AJ, Proudfoot AE, Power CA, Wells TN (1997) Effect of a CC chemokine receptor antagonist on collagen induced arthritis in DBA/1 mice. *Immunol Lett* 57: 117–120
27 Barnes DA, Tse J, Kaufhold M, Owen M, Hesselgesser J, Strieter R, Horuk R, Perez HD (1998) Polyclonal antibody directed against human RANTES ameliorates disease in the Lewis rat adjuvant-induced arthritis model. *J Clin Invest* 101: 2910–2919
28 Kasama T, Strieter RM, Lukacs NW, Lincoln PM, Burdick MD, Kunkel SL (1995) Interleukin-10 expression and chemokine regulation during the evolution of murine type II collagen-induced arthritis. *J Clin Invest* 95: 2868–2876
29 Sorensen TL, Ransohoff RM (1998) Etiology and pathogenesis of multiple sclerosis. *Sem Neurol* 18: 287–294
30 Miller DH, Khan OA, Sheremata WA, Blumhardt LD, Rice GP, Libonati MA, Willmer-Hulme AJ, Dalton CM, Miszkiel KA, O'Connor PW, International Natalizumab Multiple Sclerosis Trial G (2003) A controlled trial of natalizumab for relapsing multiple sclerosis. *N Eng J Med* 348: 15–23
31 Misu T, Onodera H, Fujihara K, Matsushima K, Yoshie O, Okita N, Takase S, Itoyama Y (2001) Chemokine receptor expression on T cells in blood and cerebrospinal fluid at relapse and remission of multiple sclerosis: imbalance of Th1/Th2-associated chemokine signaling. *J Neuroimmunol* 114: 207–212.
32 Balashov KE, Rottman JB, Weiner HL, Hancock WW (1999) CCR5(+) and CXCR3(+) T cells are increased in multiple sclerosis and their ligands MIP-1alpha and IP-10 are expressed in demyelinating brain lesions. *Proc Nat Acad Sci* 96: 6873–6878
33 Baranzini SE, Elfstrom C, Chang SY, Butunoi C, Murray R, Higuchi R, Oksenberg JR (2000) Transcriptional analysis of multiple sclerosis brain lesions reveals a complex pattern of cytokine expression. *J Immunol* 165: 6576–6582
34 Sorensen TL, Tani M, Jensen J, Pierce V, Lucchinetti C, Folcik VA, Qin S, Rottman J, Sellebjerg F, Strieter RM et al (1999) Expression of specific chemokines and chemokine receptors in the central nervous system of multiple sclerosis patients. *J Clin Invest* 103: 807–815
35 Karpus WJ, Kennedy KJ (1997) MIP-1alpha and MCP-1 differentially regulate acute

and relapsing autoimmune encephalomyelitis as well as Th1/Th2 lymphocyte differentiation. *J Leuk Biol* 62: 681–687

36 Youssef S, Wildbaum G, Karin N (1999) Prevention of experimental autoimmune encephalomyelitis by MIP-1alpha and MCP-1 naked DNA vaccines. *J Autoimmunol* 13: 21–29

37 Rottman JB, Slavin AJ, Silva R, Weiner HL, Gerard CG, Hancock WW (2000) Leukocyte recruitment during onset of experimental allergic encephalomyelitis is CCR1 dependent. *Eur J Immunol* 30: 2372–2377

38 Pattison J, Nelson PJ, Huie P, von Leuttichau I, Farshid G, Sibley RK, Krensky AM (1994) RANTES chemokine expression in cell-mediated transplant rejection of the kidney. *Lancet* 343: 209–211

39 Gao W, Topham PS, King JA, Smiley ST, Csizmadia V, Lu B, Gerard CJ, Hancock WW (2000) Targeting of the chemokine receptor CCR1 suppresses development of acute and chronic cardiac allograft rejection. *J Clin Invest* 105: 35–44

40 Uneda S, Hata H, Matsuno F, Harada N, Mitsuya Y, Kawano F, Mitsuya H (2003) Macrophage inflammatory protein-1 alpha is produced by human multiple myeloma (MM) cells and its expression correlates with bone lesions in patients with MM. *Brit J Hematol* 120: 53–55

41 Lean JM, Murphy C, Fuller K, Chambers TJ (2002) CCL9/MIP-1gamma and its receptor CCR1 are the major chemokine ligand/receptor species expressed by osteoclasts. *J Cell Biochem* 87: 386–393

42 Halks-Miller M, Schroeder ML, Haroutunian V, Moenning U, Rossi M, Achim C, Purohit D, Mahmoudi M, Horuk R (2003) CCR1 is an early and specific marker of Alzheimer's disease. *Ann Neurol* 54: 638–646

43 Elices MJ (2002) BX-471 Berlex. *Curr Opin Investig Drugs* 3: 865–869

44 Gladue RP, Zwillich SH, Clucas AT, Brown MF (2004) CCR1 antagonists for the treatment of autoimmune diseases. *Curr Opin Investig Drugs* 5: 499–504

45 Liang M, Mallari C, Rosser M, Ng HP, May K, Monahan S, Bauman JG, Islam I, Ghannam A, Buckman B et al (2000) Identification and characterization of a potent, selective, and orally active antagonist of the CC chemokine receptor-1. *J Biol Chem* 275: 19000–19008

46 Horuk R, Clayberger C, Krensky AM, Wang Z, Grone HJ, Weber C, Weber KS, Nelson PJ, May K, Rosser M et al (2001) A non-peptide functional antagonist of the CCR1 chemokine receptor is effective in rat heart transplant rejection. *J Biol Chem* 276: 4199–4204

47 Horuk R, Shurey S, Ng HP, May K, Bauman JG, Islam I, Ghannam A, Buckman B, Wei GP, Xu W et al (2001) CCR1-specific non-peptide antagonist: efficacy in a rabbit allograft rejection model. *Imm Let* 76: 193–201

48 Anders HJ, Vielhauer V, Frink M, Linde Y, Cohen CD, Blattner SM, Kretzler M, Strutz F, Mack M, Grone HJ et al (2002) A chemokine receptor CCR-1 antagonist reduces renal fibrosis after unilateral ureter ligation. *J Clin Invest* 109: 251–259

49 Brown MF, Avery M, Brissette WH, Chang JH, Colizza K, Conklyn M, DiRico AP,

Gladue RP, Kath JC, Krueger SS et al (2004) Novel CCR1 antagonists with improved metabolic stability. *Bioorg Med Chem Lett* 14: 2175–2179

50 Kath JC, Brissette WH, Brown MF, Conklyn M, DiRico AP, Dorff P, Gladue RP, Lillie BM, Lira PD, Mairs EN et al (2004) Potent small molecule CCR1 antagonists. *Bioorg Med Chem Lett* 14: 2169–2173

51 Kath JC, DiRico AP, Gladue RP, Martin WH, McElroy EB, Stock IA, Tylaska LA, Zheng D (2004) The discovery of structurally novel CCR1 antagonists derived from a hydroxyethylene peptide isostere template. *Bioorg Med Chem Lett* 14: 2163–2167

52 Haringman JJ, Kraan MC, Smeets TJ, Zwinderman KH, Tak PP (2003) Chemokine blockade and chronic inflammatory disease: proof of concept in patients with rheumatoid arthritis. *Ann Rheum Dis* 62: 715–721

53 Hesselgesser J, Ng HP, Liang M, Zheng W, May K, Bauman JG, Monahan S, Islam I, Wei GP, Ghannam A et al (1998) Identification and characterization of small molecule functional antagonists of the CCR1 chemokine receptor. *J Biol Chem* 273: 15687–15692

54 Naya A, Sagara Y, Ohwaki K, Saeki T, Ichikawa D, Iwasawa Y, Noguchi K, Ohtake N (2001) Design, synthesis, and discovery of a novel CCR1 antagonist. *J Med Chem* 44: 1429–1435

Small molecule CCR2 antagonists

Paul J. Higgins, C. Eric Schwartz and Jean-Marie Nicolas

UCB Research/UCB Pharma, 840 Memorial Drive, Cambridge, MA 02139, USA

Introduction

Monocyte chemoattractant protein-1 (MCP-1), also known as CCL2, is a member of the CC family of chemokines. During inflammatory conditions, the production of MCP-1 is upregulated in both immune and non-immune cell types including macrophages, mast cells, fibroblasts, endothelial and epithelial cells, smooth muscle cells, and astrocytes. MCP-1 is a potent activator and chemoattractant of leukocytes such as monocytes, macrophages, activated T cells, dendritic cells, and natural killer (NK) cells [1]. The biological effects of MCP-1 are mediated by interaction with its primary receptor, CCR2. In addition to inducing chemotaxis, other cellular responses induced by MCP-1 include integrin activation, inflammatory cytokine production, and histamine release [2, 3].

It is well established that MCP-1 and CCR2 are central factors in the regulation of inflammatory disease. Despite the redundancy of chemokines (e.g., MCP-2, MCP-3, MIP-1α, MIP-1β, RANTES, I-309) and chemokine receptors (CCR1, CCR5, CCR8) known to induce activity in the same cell types responsive to MCP-1 [1], inhibition of the MCP-1–CCR2 interaction can significantly reduce the severity of the inflammatory response. For example, experiments that have used either anti-MCP-1 antibodies or truncated MCP-1 to neutralize CCR2 activation have demonstrated clear inhibition of cellular influx, tissue damage, and disease symptoms in animal models of arthritis [4], EAE [5], atherosclerosis [6], lung hypersensitivity [7], and nephritis [8]. Similar results have been observed in some of these models using either MCP-1 or CCR2 knockout mice [9–11].

In humans, MCP-1 plays a key role in the pathogenesis of numerous inflammatory diseases. Clinical studies have shown blood and tissue levels of MCP-1 to be significantly elevated in patients suffering from asthma [12], chronic obstructive pulmonary disease (COPD) [13], rheumatoid arthritis (RA) [14], atherosclerosis [15], and multiple sclerosis (MS) [16]. Further, MCP-1 levels often correlate with the severity of disease symptoms. In some cases the incidence of human disease is correlated with polymorphisms of the MCP-1 gene [17].

Chemokine Biology – Basic Research and Clinical Application, Volume II
edited by Kuldeep Neote, Gordon L. Letts and Bernhard Moser

Table 1 - Reported CCR2 antagonists in drug development

Company	Compound	CCR2 IC_{50}	Indications	Status
Roche/Iconix	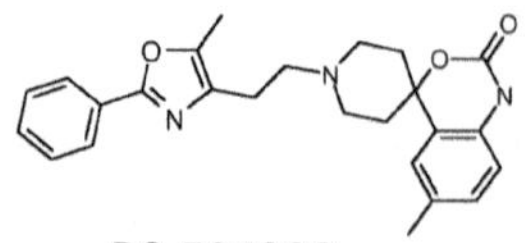 RS-504393	89 nM bind 210 nM taxis	RA, MS, athero-sclerosis	No devel-opment reported since 2000
Millennium/ Pfizer	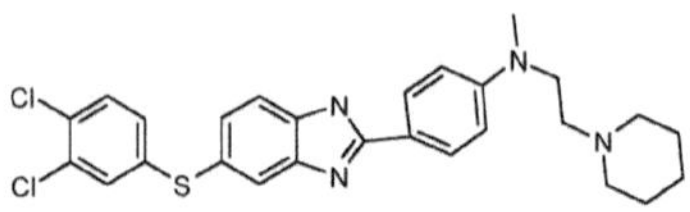 Benzimidazoles	200–300 nM bind	RA, MS, athero-sclerosis, inflammation	Discontinued
SmithKline	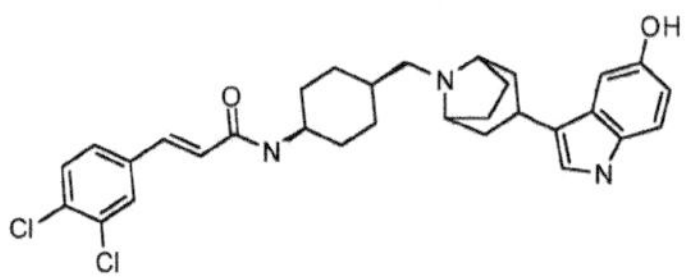 SB-380732	50 nM bind	Inflammation	Discontinued
AstraZeneca	AZD-6942	29 nM bind 60 nM taxis	RA	Discontinued
Merck	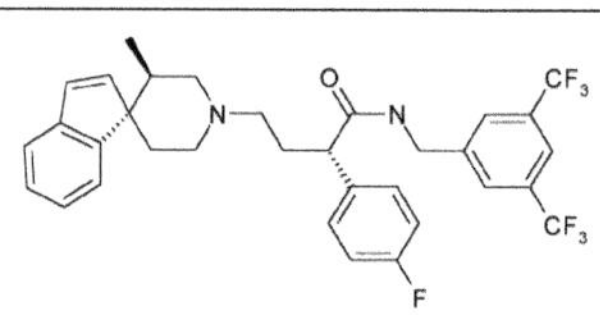	41 nM bind 59 nM taxis	RA, MS, athero-sclerosis	Preclinical
Teijin/BMS	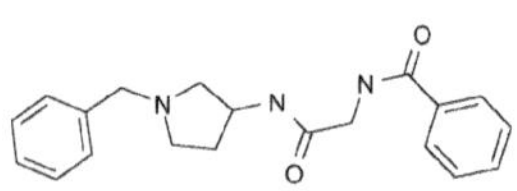 3-Aminopyrrolidines	3 nM	RA, MS, nephritis	Preclinical
Telik	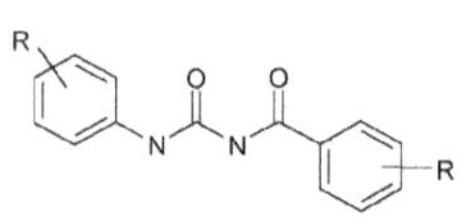		Inflamma-, tion, cancer	Preclinical
Incyte	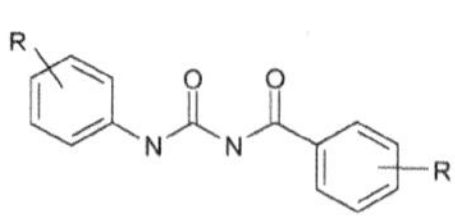 INCB-003284		Inflamma-tion	Phase I

Given the significant body of evidence implicating the role of MCP-1 in disease, CCR2 antagonism represents an attractive therapeutic strategy. Over the past several years a number of drug companies have sought to develop small molecule CCR2 antagonists. This review is intended to summarize the industry's efforts to generate CCR2 antagonist drugs with an emphasis on a description of UCB's novel compound, ucb-102405.

CCR2 antagonists in development

Dating back to 1997, over 40 patents have been published claiming non-peptide CCR2 antagonists. Various reports have indicated that a number of these compounds have entered drug development status. A list of small molecule CCR2 antagonists that have been reported as drug candidates, in addition to their stage of development, is shown in Table 1. Although specific reasons for discontinuing compound development are not clear, cross-reactivity with other G-protein coupled receptors (GPCRs) may be an issue. For example, RS-504393 interacts potently with α1 adrenergic receptors [18] and SB-380732 binds to both 5-HT and dopamine receptors [19]. The structure of AZD-6942, though not shown in drug reports, is likely the benzylindole-2-carboxylic acid scaffold that has been reported [20] and for which several patent applications have been published (e.g., WO-09907351, WO-09907678, WO-09940913, and WO-00046195). The only small molecule CCR2 antagonist known to be currently in clinical trials is Incyte's INCB-003284, which bears strong similarity to the Teijin/BMS class of compounds. The only other CCR2 antagonist in clinical trials is Millennium's anti-CCR2 neutralizing antibody, currently in Phase II for RA.

Other representative patented CCR2 antagonist scaffolds having no reported drug development activity are shown in Table 2. The Takeda compound (TAK-779), which has strong affinity to CCR5, was previously under development as a treatment for acquired immune deficiency syndrome (AIDS) but has since been discontinued.

ucb-102405

UCB Research has generated a novel patented CCR2 antagonist scaffold. One specific compound of this class, ucb-102405, has been well characterized and has exhibited the properties of a potential drug candidate. ucb-102405 potently and specifically inhibits MCP-1 binding to CCR2 as well as MCP-1-mediated biologic activities such as Ca^{2+} flux and chemotaxis (Tab. 3). However, the compound shows no *in vitro* cytotoxic effects up to 100 μM. The compound preferentially interacts with the human receptor, as it has nearly 60-fold poorer activity against murine

Table 2 - Patented CCR2 antagonists with no reported drug development activity

Company	Compound Structure	Patent #	IC_{50} CCR2	Cross-reactivity
Takeda	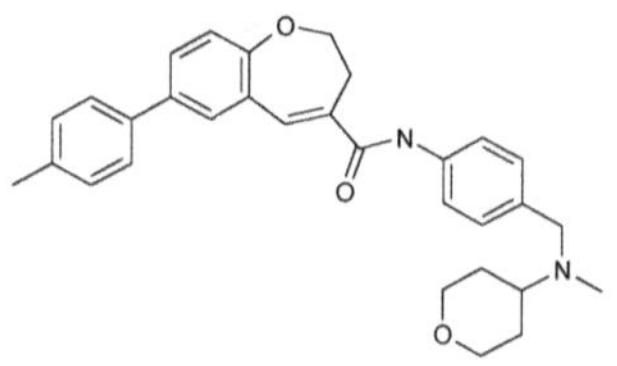	WO-09932100 WO-09932468	< 100 nM	CCR5
Chemokine Therapeutics	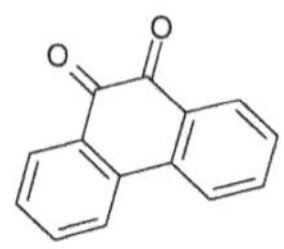	WO-00245702	10 nM	CXCR1 & 2
Pfizer	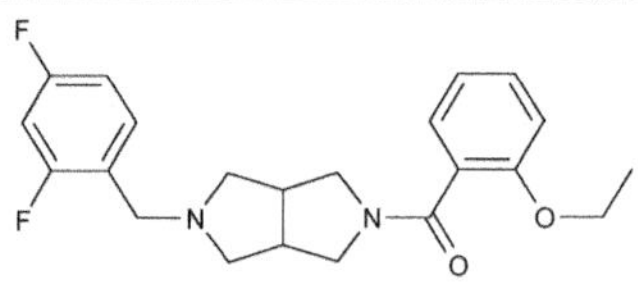	WO-02070523		CCR3
Ono	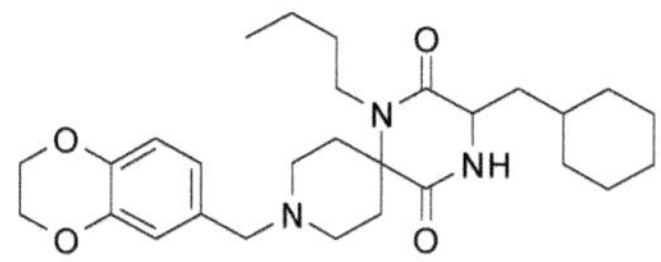	WO-02074769	40 nM	CCR5
Ono	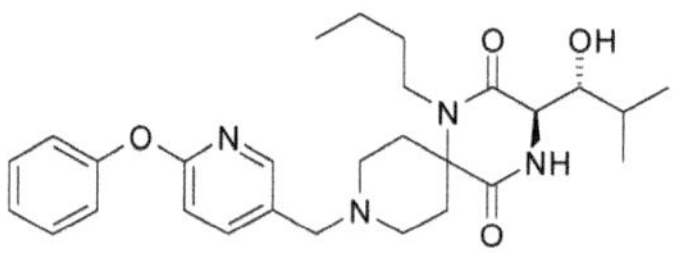	WO-02074770	27 nM	CCR5
Merck	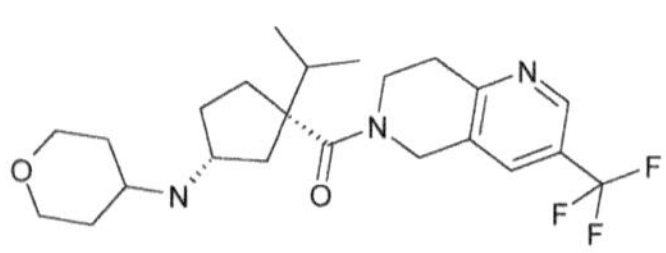	WO-03092586 WO-03093231 WO-03093266	< 1 μM	
Warner-Lambert	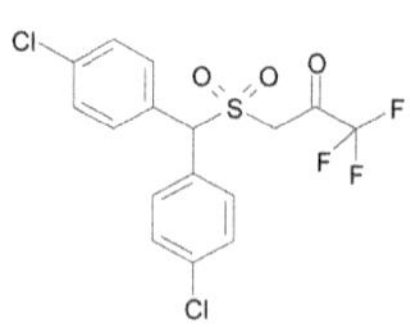	WO-2004014847	77 nM	

Table 3 - In vitro CCR2 antagonist activity of ucb-102405

Assay	Cells	IC_{50}
Binding	Human monocytes	87 nM
Ca^{2+} flux	Human THP-1 line	15 nM
Chemotaxis	Human THP-1 line	34 nM
	Human monocytes	34 nM
Cytotoxicity	Human THP-1 line	> 100 μM
Binding	Mouse WEHI line	3.7 μM

CCR2. ucb-102405 exhibits remarkably little cross-reactivity against other receptors, having no significant inhibitory activity in binding assays against 25 other GPCRs when tested up to 10 μM (data not shown). In binding assays against other chemokine receptors (CCRs 1, 3, 4, 5, 6, 7, 8 and CXCR1/2), the highest potency demonstrated was against the two most homologous receptors to CCR2, CCR1 (IC_{50} 980 nM) and CCR5 (IC_{50} 3.6 μM).

The pharmacokinetic properties of ucb-102405 have been characterized and results are shown in Figure 1. When administered at 1 mg/kg to both rats and dogs, ucb-102405 exhibited a good half-life (2–3 h), excellent oral bioavailability (> 85%), a rapid rate of absorption and high plasma levels. In addition, exposure levels in both species showed linearity with oral dose ranging from 1 mg/kg up to 200 mg/kg (data not shown).

Because MCP-1 is involved in the generation of the delayed-type hypersensitivity (DTH) response [21], this model was used to test the *in vivo* efficacy of ucb-102405. The results of experiments performed in mice are shown in Figure 2. When compound was administered 3 h after antigen challenge, a dose-dependent inhibition of skin swelling was observed. Likewise, ucb-102405 significantly inhibited the infiltration of leukocytes into the skin tissue. The high doses of compound required to induce inhibition of inflammation in mice are likely a consequence of its low cross-reactivity against murine CCR2.

Due to its potent and selective inhibition of CCR2, excellent pharmacokinetic properties and efficacy in an animal model of inflammation, ucb-102405 represents a promising candidate for development as a therapy for MCP-1-mediated disease.

Although the proof of principle for CCR2 antagonists has been demonstrated by a variety of knockout models and biological inhibitors, further experiments are needed before the efficacy of small molecule antagonists of this class is validated. Ultimately, clinical studies will determine whether CCR2 antagonists can provide effective therapy for inflammatory disease without immunosuppressive or toxic side effects.

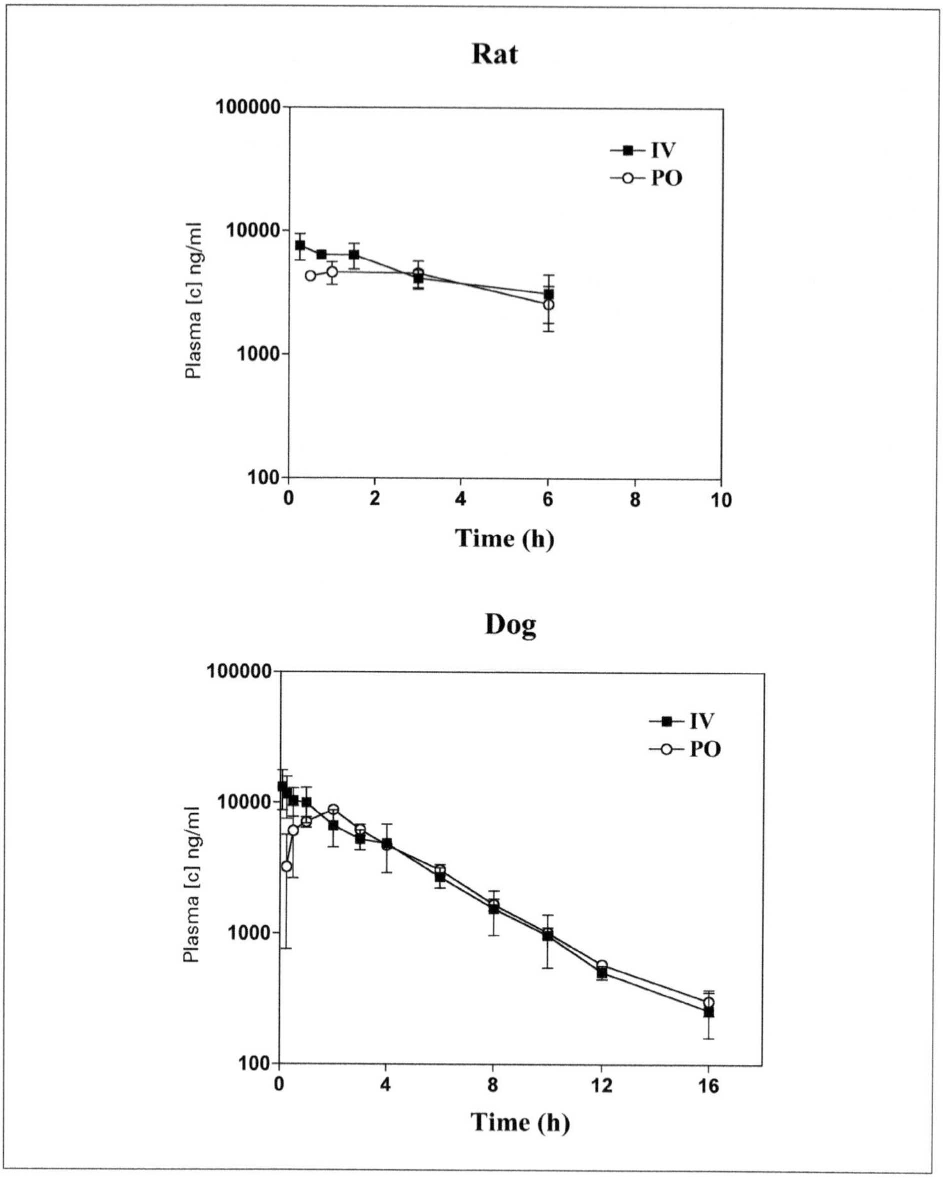

Figure 1
Pharmacokinetics of ucb-102405 in rat and dog. Male rats and dogs were administered a dose of 1 mg/kg ucb-102405 either i.v. or p.o. At various times, blood samples were drawn and plasma isolated. Concentrations of compound were determined by quantitation against a standard curve. Pharmacokinetic parameters for rat: $T_{1/2}$ *= 3.3 h, F = 93%; for dog:* $T_{1/2}$ *= 2.4 h, F = 85%.*

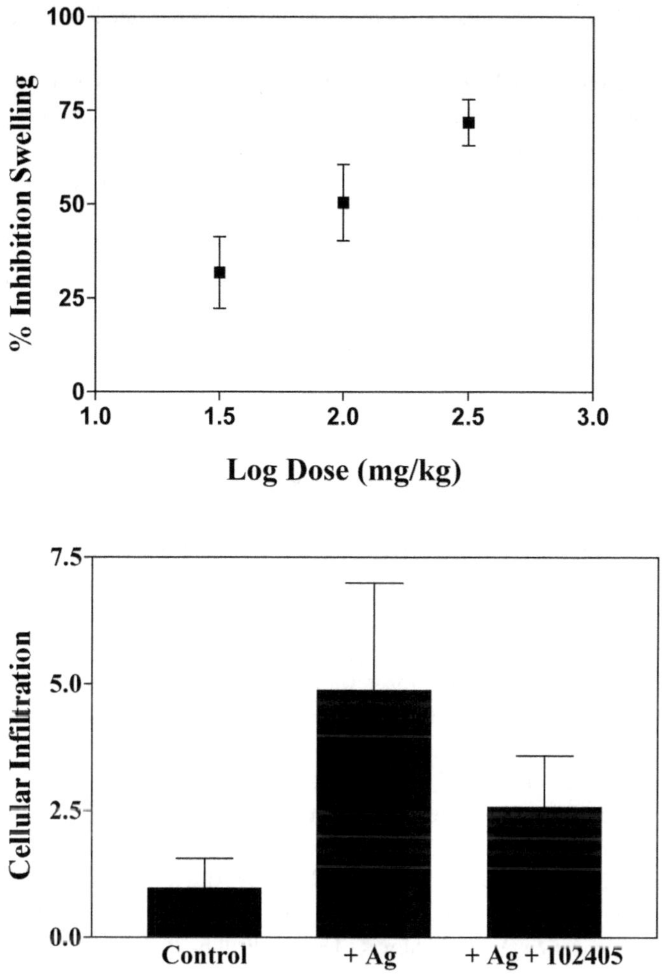

Figure 2
The effect of ucb-102405 in the DTH reaction in mice. Male Balb/c mice were immunized s.c. with ovalbumin emulsified in CFA. 7 days later, animals were challenged by i.d. injection of ovalbumin in saline into one ear. 3 h after challenge, mice were administered varying doses of ucb-102405 by i.p. injection. At 24 h after challenge ear swelling was measured by microcaliper, after which ears were removed and fixed in 1% formalin. Thin sections were prepared, stained with H & E, and image analysis performed to quantitate cell influx. The dose of ucb-102405 shown in the figure is 100 mg/kg.

References

1 Luster A (1998) Chemokines – chemotactic cytokines that mediate inflammation. *New Eng J Med* 338: 436–445

2 Biswas S, Sodhi A (2002) *In vitro* activation of murine peritoneal macrophages by MCP-1: Upregulation of CD11b, production of proinflammatory cytokines, and the signal transduction pathway. *J Interferon Cytokine Res* 22: 527–538

3 Bischoff S, Krieger M, Brunner T, Dahinden C (1992) MCP-1 is a potent activator of human basophils. *J Exp Med* 175: 1271–1275

4 Ogata H, Takeya M, Yoshimura T, Takagi K, Takahashi K (1997) The role of MCP-1 in the pathogenesis of collagen-induced arthritis in rats. *J Pathol* 182: 106–114

5 Kennedy K, Strieter R, Kunkel S, Lukacs N, Karpus W (1998) Acute and relapsing EAE are regulated by differential expression of the CC chemokines MIP-1α and MCP-1. *J Neuroimmunol* 92: 98–108

6 Inoue S, Egashira K, Ni W, Kitamoto S, Usui M, Otani K, Ishibashi M, Hiasa K, Nishida K, Takeshita A (2002) Anti-MCP-1 gene therapy limits progression and destabilization of established atherosclerosis in apolipoprotein E knockout mice. *Circulation* 106: 2700–2706

7 Gonzalo J, Lloyd C, Wen D, Albar J, Wells T, Proudfoot A, Martinez-A C, Dorf M, Bjerke T, Coyle A et al (1998) The coordinated action of CC chemokines in the lung orchestrates allergic inflammation and airway hyperresponsiveness. *J Exp Med* 188: 157–167

8 Wolf G, Jocks T, Zahner G, Panzer U, Stahl R (2002) Existence of a regulatory loop between MCP-1 and TGF-β in glomerular immune injury. *Am J Physiol* 283: F1075–F1084

9 Campbell E, Charo I, Kunkel S, Strieter R, Boring L, Gosling J, Lukacs N (1999) MCP-1 mediates cockroach allergen-induced bronchial hyperreactivity in normal but not $CCR2^{-/-}$ mice: The role of mast cells. *J Immunol* 163: 2160–2167

10 Dawson T, Kuziel W, Osahar T, Maeda N (1999) Absence of CCR2 reduces atherosclerosis in apolipoprotein E-deficient mice. *Atherosclerosis* 143: 205–211

11 Tesch G, Schwarting A, Kinoshita K, Lan H, Rollins B, Kelley V (1999) MCP-1 promotes macrophage-mediated tubular injury, but not glomerular injury, in nephrotoxic serum nephritis. *J Clin Invest* 103: 73–80

12 Alam R, York J, Boyars M, Stafford S, Grant J, Lee J, Forsythe P, Sim T, Ida N (1996) Increased MCP-1, RANTES, and MIP-1α in bronchoalveolar lavage fluid of allergic asthmatic patients. *Am J Respir Crit Care Med* 153: 1398–1404

13 Jahnz-Rozyk K, Kuna P, Pirozynska E (1997) MCP-1 in bronchoalveolar lavage fluid from patients with atopic asthma and chronic bronchitis. *J Invest Allergol Clin Immunol* 7: 254–259

14 Ellingsen T, Buus A, Stengaard-Pederson K (2001) Plasma MCP-1 is a marker for joint inflammation in rheumatoid arthritis. *J Rheumatol* 28: 41–46

15 Nelken N, Coughlin S, Gordon D, Wilcox J (1991) MCP-1 in human atheromatous plaques. *J Clin Invest* 88: 1121–1127
16 Simpson J, Newcombe J, Cuzner M, Woodroofe M (1998) Expression of MCP-1 and other β-chemokines by resident glia and inflammatory cells in multiple sclerosis lesions. *J Neuroimmunol* 84: 238–249
17 Szalai C, Kozma G, Nagy A, Bojszko A, Krikovszky D, Szabo T, Falus A (2001) Polymorphism in the gene regulatory region of MCP-1 is associated with asthma susceptibility and severity. *J Allergy Clin Immunol* 108: 375
18 Mirzadegan T, Diehl F, Ebi B, Bhakta S, Polsky I, McCarley D, Mulkins M, Weatherhead G, Lapierre J, Dankwardt J et al (2000) Identification of the binding site for a novel class of CCR2b chemokine receptor antagonists. *J Biol Chem* 275: 25562–25571
19 Witherington J, Bordas V, Cooper D, Forbes I, Gribble A, Ife R, Berkhout T, Gohil J, Groot P (2001) Conformationally restricted indolopiperidine derivatives as potent CCR2b receptor antagonists. *Bioorg Med Chem Lett* 11: 2177–2180
20 Kettle J, Faull A, Barker A, Davies D, Stone M (2004) N-Benzylindone-2-carboxylic acids: potent functional antagonists of the CCR2b chemokine receptor. *Bioorg Med Chem Lett* 14: 405–408
21 Rand M, Warren J, Mansour M, Newman W, Ringler D (1996) Inhibition of T cell recruitment and cutaneous delayed-type hypersensitivity-induced inflammation with antibodies to MCP-1. *Am J Pathol* 148: 855–864

Chemokine axes in hematopoietic stem cell mobilization

Louis M. Pelus and Hal E. Broxmeyer

Department of Microbiology & Immunology and the Walther Oncology Center, Indiana University School of Medicine and the Walther Cancer Institute, 950 West Walnut Street, Indianapolis, IN 46202, USA

Introduction: Hematopoietic stem cell mobilization

Under steady state, hematopoietic cells at all stages of differentiation and some lymphoid cell populations are confined within the bone marrow, while mature cells exit and enter peripheral blood. A small population of hematopoietic stem (HSC) and progenitor (HPC) cells also traffic through the peripheral circulation. The initial observations that HPC cells are found in the blood of patients recovering from chemotherapy [1–3] led to the realization that HSC and HPC can be forced or 'mobilized' from marrow to blood where they can be collected by apheresis and utilized for transplantation [4].

The clinical utility of mobilized hematopoietic grafts is now well established and mobilized peripheral blood stem cells (PBSC) have become the primary graft used to reconstitute hematopoiesis for autologous and allogeneic transplantation. Although hematopoietic stem cell mobilization was first described as a consequence of chemotherapy, it required robust doses and was often associated with complications of neutropenic sepsis, bleeding and rarely even death [4]. In addition, chemotherapy-induced mobilization was found to be unpredictable [4] and of short duration [5]. We now know that many cytokines and other agents can mobilize hematopoietic stem and progenitor cells (reviewed in [4, 6–8]), many of which are additive or synergistic when used in combination. Clinically, granulocyte colony-stimulating factor (G-CSF) has emerged as the predominant peripheral blood stem cell mobilizer based upon potency, predictability and safety. G-CSF-mobilized PBSC exhibit more rapid engraftment compared to bone marrow or umbilical cord blood [4, 9–11], resulting in faster neutrophil and platelet recovery, fewer platelet transfusions, faster lymphocyte reconstitution and fewer febrile episodes [4, 12–15]. Overall, PBSC collection is less invasive than marrow harvest, avoiding risks associated with general anesthesia. Target yields are generally achieved with 1–3 aphereses, which is usually adequate for human leukocyte antigen (HLA) identical and matched related transplant, but is often suboptimal in the haploidentical setting [16–18].

Although widely used, inadequate/suboptimal mobilization in response to G-CSF occurs in 25% of patients, particularly those with lymphomas and myelomas [19].

Chemokine Biology – Basic Research and Clinical Application, Volume II
edited by Kuldeep Neote, Gordon L. Letts and Bernhard Moser

Patients with Fanconi anemia also respond poorly to mobilization by G-CSF [20]. For allogeneic transplant, 10–20% of normal volunteers do not respond well to G-CSF for unknown reasons [21], requiring extended aphereses to collect sufficient cells for transplant [22]. Expanded application of allogeneic transplantation, particularly to patients with malignant disease receiving grafts from matched unrelated donors, to patients with non-malignant disease receiving grafts to induce tolerance for subsequent solid organ transplant, or for treatment of genetic disorders of hematopoiesis or inborn errors of metabolism that do not require lethal conditioning, will require larger doses of mobilized cells to overcome HLA barriers and provide durable engraftment, acceptable leukocyte recovery kinetics and low incidence of graft *versus* host disease (GVHD) [16, 17]. Thus despite the success of G-CSF, there are areas for improvement that can have significant clinical impact. These include the requirement for multiple daily injections; variable and suboptimal mobilization in subpopulations of patients and normal donors; number of aphereses procedures required to collect optimal cell numbers and inability to predict optimal mobilization times.

The chemokine axis

The chemokines are a large family of cytokines possessing the ability to direct the movement of various leukocyte populations, including hematopoietic stem cell and progenitor cells, and are involved in regulating hematopoietic responses including survival [23–27], proliferation [23, 25, 27, 28] and migration and homing [29–34]. Chemokines also influence the survival and chemotaxis of murine embryonic stem cells and production of primitive and definitive hematopoietic progenitor cells [35]. The ability of chemokines to influence the migration of hematopoietic stem and progenitor cells led to analysis of their abilities as stem cell mobilizers or stem cell retention factors and the realization that a number of chemokines can mobilize hematopoietic populations. In general, the hallmarks of chemokine mobilization are rapid mobilization, with peak responses occurring in minutes and hours rather than the multiple days required for G-CSF or other growth factors that induce mobilization responses (reviewed in Tab. 1) and enhanced or synergistic mobilization responses when combined with other cytokines, particularly G-CSF (reviewed in Tab. 2). Chemokines, as a family of cytokines, are characterized by redundancy, and the role of these chemotactic factors in mobilization is not an exception, at least at first look.

Macrophage inflammatory protein-1α (MIP-1α)

The first chemokine shown to mobilize hematopoietic cells was macrophage inflammatory protein-1 (MIP-1α/CCL3), which serves as a ligand for both the CCR1 and

Table 1 - Mobilization of hematopoietic stem and progenitor cells in mice by chemokines or agents that influence chemokine/chemokine receptor interactions compared to G-CSF

Mobilizer	Timing of response	Fold increase over baseline		Long-term repopulation
		CFU-GM/ml blood	SKL cells/ml blood[a]	
rh-G-CSF	50 ug/kg;SC; bid x 4 days	24.4 [43]	11.9[b]	(++)[c]
rhIL-8/CXCL8	15 min post 1 mg/kg; SC	4.6 [43]	–	(++)
rhMIP1-α/CCL3 (BB10010)	15 min post 1 mg/kg; SC	3.7 [43]	–	ND
rhGROβ/CXCL2	15 min post 1 mg/kg; SC	11.4 [43]	–	(++)
rhGROβ$_{\Delta 4}$ (SB251353)	15 min post 1 mg/kg; SC	11.7 [43]	11.5[b]	(++)
rmMIP2	15 min post 20 ug/kg; IV	–	7.6 [46]	ND
SDF-1β/CXCL12	24 hrs post 15 mg/kg; IV	–	0.9 [63][d]	ND
Met-SDF-1β/ CXCL12	24 hrs post 15 mg/kg; IV	–	32.5 [63][d]	ND
AdSDF-1α/ CXCL12	day 5 post Adenovirus; IV	15 [64]	18 [64][e]	(++)
rhSDF-1α/ CXCL12	60 min post 1 mg/kg; IV	1.8 [78]	–	ND
CTCE-0021	60 min post 25 mg/kg; IV	14.3 [80]	5.82 [78][f]	ND
CTCE-0214	60 min post 25 mg/kg; IV	22.1 [80]	2.9 [81]	ND
rh-G-CSF	50 ug/kg;SC;bid x 4 days	36.4 [80]	–	ND
AMD3100	60 min post 5 mg/kg; SC	7.1 [88]	–	(++)

[a]*SKL = Sca-1$^+$, c-kit$^+$, lineage$^-$ cells*
[b]*L Pelus, unpublished*
[c]*(++) denotes engraftment activity in lethally irradiated mice*
[d]*SKL-Thy^{lo}*
[e]*Sca-1$^+$ only*
[f]*Kit$^+$, Lin$^-$ only*
ND = No data available

CCR5 receptors. Murine studies with a genetically engineered variant of MIP-1α/CCL3, BB10010, demonstrated rapid mobilization of neutrophils, lineage restricted progenitor cells and more primitive progenitor cell populations known to contain short-term marrow repopulating cells [36, 37]. Modest four-fold enhanced mobilization was observed in combination with G-CSF. In a Phase I clinical trail in women with Stage IIIB or IV breast cancer not having received prior cyclophos-

Table 2 - Mobilization of hematopoietic stem and progenitor cells in mice by the combination of chemokines or agents that influence chemokine/chemokine receptor interactions and G-CSF

Mobilizer	G-CSF Regimen	Timing of response[a]	Combination with G-CSF			
			Fold increase over baseline		Fold increase over G-CSF	
			CFU-GM/ml blood	SKL cells/ml blood[b]	CFU-GM	SKL cells
rhIL-8/CXCL8	125 ug/kg; bid x 2d	15 min post 1 mg/kg	40 [41]	–	8 [41]	–
rhMIP1-α/CCL3	125 ug/kg; bid x 2d	15 min post 1 mg/kg	38.8[c] [36]	–	1.5[c] [36]	–
rhGROβ/CXCL2	50 ug/kg; bid x 4d	15 min post 1 mg/kg	24.7 [45]	–	5.5 [45]	–
rhGROα$_{\Delta 4}$/CXCL2$_{\Delta 4}$	50 ug/kg; bid x 4d	15 min post 1 mg/kg	26.8 [45]	29[d]	5.6 [45]	2.7[d]
rmMIP2	125 ug/kg; bid x 2d	15 min post 20 ug/kg	–	19 [46]	–	2.1 [46]
CTCE-0021	50 ug/kg; bid x 4d	60 min post 25 mg/kg	190 [80]	–	5.2 [80]	–
CTCE-0214	50 ug/kg; bid x 4d	60 min post 25 mg/kg	247 [80]	–	7.1 [80]	–
rh-G-CSF	50 ug/kg; bid x 4d	–	24–36[e]	–	–	–
AMD3100	125 ug/kg; bid x 2d	60 min post 5 mg/kg	51.1 [88]	–	5.3 [88]	–

[a]*Chemokines administered at ~16 hours post last injection of G-CSF*
[b]*SKL = Sca-1$^+$, c-kit$^+$, lineage$^-$ cells*
[c]*Day 8 CFU-S*
[d]*L Pelus, unpublished*
[e]*Fold increase data for G-CSF at baseline is shown in Table 1*

phamide and not more than one chemotherapy regimen for metastatic disease, BB10010 demonstrated a significant but modest mobilizing effect on myeloid progenitor cells ranging from 3.4- to 5.6-fold [37], with little differences noted at doses between 5–100 ug/kg. Maximal mobilization was usually observed after 3 days of dosing, although in some cases, maximal responses were observed within 24 h, which are significantly longer than observed in mice. The relatively modest mobilizing capacity in patients with breast cancer suggests that clinical utility would require combination use with other cytokines such as G-CSF. Mechanistically, CCR1 is clearly a dominant receptor for MIP-1α-induced mobilization since mice in which the CCR1 gene is deleted do not mobilize in response to MIP-1α [38]. Interestingly, CCR1 knockout mice are more responsive to mobilization by G-CSF than wild-type littermate controls [38], suggesting that CCR1 and its ligand(s) may play a negative or dampening role on G-CSF-induced PBSC mobilization, at least in the mouse.

CXCR2 ligands: Interleukin-8 (IL-8), GROβ and macrophage inflammatory protein-2 (MIP-2)

A number of ligands for the CXCR2 receptor, including IL-8/CXCL8 [39–41], the GRO proteins, particular GROβ/CXCL2 and a 4 amino truncated variant termed $GRO\beta_{\Delta 4}/CXCL2_{\Delta 4}$ (SB251353) [42–45] and murine MIP-2 (KC), the murine homolog of IL-8 and/or GROβ [46], have shown activity as rapid peripheral blood stem cell mobilizers in mice and monkeys.

IL-8/CXCL8 is a proinflammatory CXC chemokine produced by a variety of cells including monocytes, neutrophils, endothelial cells, epithelium, at least in the lung, mast cells and keratinocytes. It binds to both the CXCR1 (IL-8 specific) and CXCR2 (IL-8 permissive) receptors, and is responsible for a variety of biological effects including stimulation of neutrophil chemotaxis and matrix metalloproteinase-9 (MMP-9) release (reviewed in [47]) and inhibition of hematopoietic progenitor cell proliferation [48]. It also induces shedding of L-selectin and upregulation of the β2-integrin, LFA-1 [40]. In mice, injection of IL-8/CXCL8 alone produces a rapid neutropenia followed by neutrophilia several hours later. IL-8/CXCL8 also produces a rapid but transient mobilization of hematopoietic progenitor cells and stem cells with radioprotective and long-term reconstitution ability, which peak within 15 min and disappear within 1 h [39]. IL-8/CXCL8 has not been clinically evaluated; however in rhesus monkeys it induces a similar rapid release of HPC capable of colony formation *in vitro* [49]. Combination studies using IL-8/CXCL8 and other mobilizing cytokines such as G-CSF have not been performed in monkeys. Addition of IL-8/CXCL8 to mice pretreated with either G-CSF or SCF resulted in enhanced HPC mobilization compared to any of the agents used alone [41]. A prominent finding in these studies was a role for polymorphonuclear neutrophils

(PMN) in the mobilization process, in that PBSC mobilization induced by IL-8/CXCL8 was absent in mice made neutropenic by an anti-PMN specific antibody and recovered concomitant with neutrophil recovery [50]. A role for neutrophils in IL-8/CXCL8 mobilization is supported by the findings that mobilization in mice and rhesus monkeys could be blocked by anti-LFA-1 [51] and anti-MMP-9 antibodies [52], respectively. Neutrophils express LFA-1 [53] and PMN granules are a major storage site of MMP-9 [54]. Antibodies to β2 integrins likely block IL-8/CXCL8 induced mobilization by blocking MMP-9 release from PMN rather than from inhibition of ligand binding to ICAM-1, since antibodies to β2 integrins alone do not induce mobilization [55, 56].

The GROβ proteins are selective ligands for the CXCR2 (IL-8 permissive) receptor and unlike IL-8 do not bind to the CXCR1 receptor. GROβ/CXCL2 is primarily produced by fibroblasts and stromal cells. The N-terminal four amino acid truncated variant of GROβ/CXCL2, termed $GRO\beta_{\Delta4}/CXCL2_{\Delta4}$, results from peptidase processing. Unlike IL-8/CXCL8, the GROβ proteins do not have significant proinflammatory activities. GROβ/CXCL2 and $GRO\beta_{\Delta4}/CXCL2_{\Delta4}$ stimulate neutrophil respiratory burst activity, CD11b expression and killing of opportunistic pathogens, with the truncated variant being significantly more potent than the full length protein [44]. Both GRO proteins rapidly mobilize HSC, including all classes of short-term progenitor cells as well as long-term repopulating stem cells (LTRC) [42]. In mice, peak GROβ/CXCL2 or $GRO\beta_{\Delta4}/CXCL2_{\Delta4}$ induced mobilization occurs within 15–30 min and does not generally persist beyond 60 min. In rhesus monkeys, peak mobilization occurs within 1 h and persists for more than 4 h, which is expected based upon pharmacokinetics. Similar to IL-8/CXCL8, PBSC mobilization induced by GROβ/CXCL2 and $GRO\beta_{\Delta4}/CXCL2_{\Delta4}$ require polymorphonuclear neutrophils and neutrophil derived active MMP-9 [45]. Mobilization in MMP-9 knockout mice is significantly reduced [45], although not absent, which implies possible redundancies in MMP-9 activity. In this regard, the MMPs are a large family of endopeptidases [57–59], with each MMP having both preferred and overlapping substrate specificities [58, 60]. In monkeys and/or mice, mobilization by single administration of $GRO\beta_{\Delta4}/CXCL2_{\Delta4}$ is similar in magnitude to a multiday regimen of G-CSF and both GROβ proteins demonstrate synergistic mobilization of HPC and long-term repopulating stem cells when used in combination with G-CSF. In rhesus monkeys, both granulocyte-macrophage (CFU-GM) and megakaryocyte (CFU-Meg) (58-fold increase at 45 min post 250 ug/kg, *sc*) progenitor cells were significantly mobilized by a single *sc* injection of $GRO\beta_{\Delta4}/CXCL2_{\Delta4}$ in a dose response manner [42]. In contrast, IL8/CXCL8 is a poor mobilizer of CFU-Meg in rhesus monkeys [49], clearly suggesting differences in CXC chemokine mobilization. The ability of $GRO\beta_{\Delta4}/CXCL2_{\Delta4}$ to induce CFU-GM mobilization in monkeys, i.e., 11.2-fold following a single dose of 500 ug/kg, was equivalent to G-CSF, which induced an 11.4-fold increase with a regimen of 10 ug/kg/day, SC, x 4 days [42].

The combination of G-CSF plus GRO$\beta_{\Delta4}$/CXCL2$_{\Delta4}$ augmented CFU-GM and CFU-Meg mobilization above that observed for G-CSF alone [42]. In two monkeys treated, G-CSF administration resulted in 17- and 24-fold increases in CFU/ml of blood after 4 days. The addition of 250 ug/kg GRO$\beta_{\Delta4}$/CXCL2$_{\Delta4}$ 24 h after the last dose of G-CSF boosted these levels to 26.1- and 26.7-fold over baseline, representing a mean two-fold increase in CFU-GM and a 200-fold increase in CFU-Meg over that observed with G-CSF alone. This is highly significant, realizing that the dose of GROβΔ4/CXCL2Δ4 used in these studies was less than optimal. Mobilization of CFU-Meg was not routinely observed following administration of G-CSF in these rhesus monkey studies. In mice, combination mobilization with GROβ/CXCL2 or GRO$\beta_{\Delta4}$/CXCL2$_{\Delta4}$ and G-CSF is highly synergistic compared to either agent used alone, and directly correlates with the presence of neutrophils and PMN-derived active MMP-9 [45]. In addition, synergistic combination mobilization is completely blocked in MMP-9 knockout mice, indicating that the potential compensatory mechanisms that allow for some degree of mobilization in response to GRO$\beta_{\Delta4}$/CXCL2$_{\Delta4}$ used alone cannot substitute for MMP-9 in the setting of combination mobilization with G-CSF, supporting a primary role for MMP-9 in the synergistic mobilization response.

Transplantation of equivalent numbers of PBSC mobilized by GROβ/CXCL2 or GRO$\beta_{\Delta4}$/CXCL2$_{\Delta4}$ into lethally irradiated mice results in faster neutrophil and platelet recoveries compared to PBSC mobilized by G-CSF, which is accelerated even further when PBSC are obtained from mice mobilized by either GROβ protein used in combination with G-CSF [42, 43, 61]. In competitive repopulation transplant experiments, GRO$\beta_{\Delta4}$/CXCL2$_{\Delta4}$ and G-CSF plus GRO$\beta_{\Delta4}$/CXCL2$_{\Delta4}$ mobilized PBSC show a significant repopulation advantage compared to normal bone marrow cells or PBSC mobilized by G-CSF. Secondary transplantation analysis demonstrated that PBSC mobilized by GRO$\beta_{\Delta4}$/CXCL2$_{\Delta4}$ alone or in combination with G-CSF produce durable engraftment [45]. In addition, mobilization by a single dose of either GROβ protein with only 1 day's administration of G-CSF is equivalent in magnitude to a 4-day regimen of G-CSF [42, 43]. These studies demonstrate that the GROβ proteins are rapid and efficacious mobilizers when used alone, and synergistic when used in combination with G-CSF, which offers potential new paradigms for hematopoietic stem and progenitor cell mobilization and enhanced clinical effectiveness of transplant regimens. They also suggest that the stem cell graft mobilized has enhanced engraftment capacity compared to cells mobilized by G-CSF. GRO$\beta_{\Delta4}$/CXCL2$_{\Delta4}$ (SB-251353) has completed Phase II clinical trials with efficacy and safety alone and in combination with G-CSF (GlaxoSmithKline, personal communication), consistent with its effects in murine and rhesus studies.

Consistent with findings using IL-8/CXCL8 or GROβ/CXCL2, the mouse homolog of GROβ, MIP-2 or KC, induces a rapid mobilization of hematopoietic progenitor cells, and Sca-1^{+}, kit^{+}, Lin^{-} (SKL) cells, known to contain a population of long-term repopulating cells [46] in mice. MIP-2 also synergistically mobilized hematopoietic cells in combination with G-CSF. The dose of MIP-2 required for mobilization was significantly lower than that of IL-8 or GROβ, indicating species specificity of the proteins, a fact also borne out in rhesus monkey studies [42]. Mobilization by the combination of MIP-2 and G-CSF was also associated with downregulation of L-selectin on KSL cells in peripheral blood and bone marrow.

Stromal cell derived factor-1α (SDF-1α)/CXCL12

The chemokine SDF-1α/CXCL12 is produced by stromal cells and its unique receptor CXCR4 is expressed on hematopoietic stem and progenitor cells, and the SDF-1/CXCL12/CXCR4 axis is believed to be involved in HSC homing and mobilization [32, 34, 62–64]. Targeted gene knockouts of either ligand or receptor result in identical disruption of hematopoiesis [65–67]. SDF-1/CXCL12 was also shown to enhance activity of adhesion receptors on hematopoietic stem and progenitor cells [68–70], serving to retain these cells within the marrow. Moreover, G-CSF mobilized cells demonstrated reduced migratory response to SDF-1 [29]. These findings led to the hypothesis that altering the SDF-1/CXCL12 gradient between peripheral blood and marrow would influence the egress or retention of HSC and HPC.

In an early study, mobilization of HPC and cells with repopulating activity was observed 5 days after adenoviral-vector delivery of human (h)SDF-1/CXCL12 [64]. Mobilization occurred coincident with a peak in plasma but not marrow SDF-1/CXCL12 [64], suggesting that altering the SDF-1 gradient between marrow and blood was responsible for mobilization. A correlation between reduction in marrow SDF-1/CXCL12 as a result of proteolysis by neutrophil elastase (NE) and mobilization by G-CSF was subsequently reported [71, 72], which also suggested that altered blood/marrow SDF-1/CXCL12 gradients might be responsible for mobilization. This attractive, although unproven, mechanism was further supported by the facts that SDF-1/CXCL12 can be processed and inactivated by NE [71], Cathepsin G (CG) [73] and MMP-9 [74], which is consistent with a primary role for PMN in the mobilization process, and a correlation between PMN mobilization and altered marrow to plasma SDF-1/CXCL12 concentration [75]. Reduction in SDF-1/CXCL12 mRNA in cytokine mobilized mice has also been shown [76, 77]. However, direct administration of SDF-1α or SDF-1β in mice failed to induce mobilization [63, 78], which was believed to reflect short serum half life and proteolytic processing that inactivated its function [73, 74, 79]. Rapid HPC mobilization by more stable cyclized lactam SDF-1α peptide analogs, CTCE-0021 and CTCE-0214, alone, and/or synergy in combination with G-CSF has recently been demonstrated [78, 80]. Mobilization by CTCE-0021 was found to be protease independent and associated with higher plasma SDF-1/CXCL12 levels, which likely represents detection of CTCE-0021 in the SDF-1 ELISA [78]. Since CTCE-0021 chemoattracts hematopoietic cells [78, 80, 81], altered gradient may contribute to its mobilizing mechanism.

Although reduced levels of SDF-1/CXCL12 are observed in the marrow of G-CSF-mobilized mice [71]; SDF-1/CXCL12 protein values in plasma measured in ngs/ml was compared to ngs per 1 ml marrow extracts, which does not truly reflect the volume of the marrow compartment in a mouse femur or the concentration of SDF-1/CXCL12 in the extravascular space. Extensive analysis comparing molarity rather than absolute levels of SDF/CXCL12 between these compartments do not support a significant shift in gradient as a mechanism for stem cell mobilization [45], although all such measurements can be criticized for not taking into account local gradients due to cell bound SDF-1/CXCL12. In addition, the ELISAs used to measure SDF-1/CXCL12 do not discriminate between full length active and truncated SDF-1. This becomes important within the context of recent studies showing that CD26-mediated N-terminal processing of SDF-1 may generate SDF-1 peptides that lose the capacity to act as chemotactic agents, but are functional antagonists at CXCR4 for HPC migration [82, 83]. In contrast, hematopoietic progenitor cells from CD26 knockout mice, or mice that had CD26 inhibited by small peptides, have an attenuated mobilization response to G-CSF [82, 84].

Agents that antagonize the SDF-1/CXCL12/CXCR4 axis

AMD3100

AMD3100 is a low molecular weight bicyclan that specifically and reversibly blocks SDF-1/CXCL12 binding to and signaling through CXCR4 [85–87]. It induces rapid mobilization of hematopoietic progenitor cells, including CFU-GM, BFU-E and CFU-GEMM in mice and man and $CD34^+$ cells (that contain hematopoietic stem and progenitor cells) in man [88–92]. In mice, mobilization peaks at 1 h post administration [88], whereas in man, peak mobilization occurs between 6 and 9 h post administration, depending upon the dose used [89, 90]. Combination of a single dose of AMD3100 following 4 or 5 days administration of G-CSF, synergistically augmented G-CSF-induced mobilization in mice [88] and man [88, 92]. AMD3100 effectively mobilizes murine long-term repopulating stem cells that engraft in lethally irradiated mice, as well as human $CD34^+$ cells that repopulate immunodeficient mice (non-obese diabetic-severe combined immunodeficiency mice (NOD/SCID)). The human SCID repopulating cells are believed to represent functional human hematopoietic stem cells. AMD3100 synergizes with G-CSF to mobilize murine LTRC and human SRC [88]. Of interest, human $CD34^+$ cells isolated after treatment with G-CSF plus AMD3100 expressed a phenotype characteristic of highly engrafting mouse hematopoietic stem cells [88]. AMD3100 is currently in multicenter Phase III clinical trials as an adjunct to G-CSF-induced mobilization of hematopoietic peripheral blood stem cells for transplantation. Thus far, administration of AMD3100 to patients and normal volunteers appears to be safe.

Others

At first pass, the facts that CXCR4 agonists and antagonists both lead to mobilization seem contradictory. However, the SDF-1 analog Met-SDF-1β that is resistant to proteolytic cleavage, also stimulates a more prolonged desensitization of CXCR4 than native SDF-1 [63, 93]. Significantly lower CXCR4 expression is also found on c-kit^{+}, lineage^{negative} (KL) cells in marrow and blood of CTCE-0021 mobilized mice compared to resident cells and incubation of KL cells with CTCE-0021 confirmed its ability to downregulate CXCR4 by >two-fold compared to SDF-1/CXCL12 [78]. The use of antibodies selective for different regions of the extracellular domains of CXCR4 ruled out inactivating N-terminal CXCR4 cleavage. These studies in combination with the mobilizing activity of CXCR4 antagonist AMD3100, strongly suggest that altering CXCR4 signaling is involved in the mobilization response. It is of potential relevance that transforming growth factor-β (TGF-β) blocks SDF-1/CXCL12 desensitization of CXCR4 intracellular signaling, chemotaxis and adhesion of CD34^{+} cells [94], suggesting a role for TGF-β in chemokine induced mobilization. In mice receiving adenoviral delivered hSDF-1/CXCL12, CXCR4 expression was not measured; however, since chemokine receptors are rapidly internalized by ligand binding [95], mobilization could have resulted from CXCR4 desensitization due to sustained high plasma SDF-1/CXCL12 concentration. Mobilization by Pertussis toxin (PTX), a Gαi inhibitor, in mice, is consistent with mobilization by antagonizing CXCR4, which is coupled to Gαi, and a requirement for active signaling through CXCR4 for retention of cells in marrow. Moreover, PTX demonstrated mobilization synergy with G-CSF that was MMP-9 dependent, supporting studies that synergistic PBSC mobilization by GROβ plus G-CSF [45] is MMP-9 dependent. In contrast, neutralizing anti-CXCR4 antibody was reported to block mobilization by G-CSF [71], which is opposite to what one might expect, however, antibody/receptor binding is not *apriori* equivalent to receptor antagonism or desensitization.

The absence of mobilization by IL-8/CXCL8 or G-CSF in G-CSF receptor deficient mice [96], by G-CSF, IL-8/CXCL8, GROβ/CXCL2 or GRO$\beta_{\Delta 4}$/CXCL$2_{\Delta 4}$ in PMN depleted mice [45, 50], and by GRO$\beta_{\Delta 4}$/CXCL$2_{\Delta 4}$ in CXCR2 knockout mice [43], clearly establish that a functional pool of CXCR2 receptor positive neutrophils is required for stem cell mobilization. The realization that PMN granules contain a variety of proteases able to degrade components of the extracellular matrix that constitute the hematopoietic niche, including VCAM-1, CXCR4/SDF-1/CXCL12 and SCF, has implicated neutrophil-derived proteases in the mobilization cascade, although lack of mobilization by G-CSF in protease deficient mice [97] is difficult to explain at present. In addition to elevated neutrophil elastase and cathepsin G in the marrow of mice that correlates with reduction in VCAM-1 and CFU-GM mobilization in G-CSF mobilized mice [98, 99], mobilization of HPC by G-CSF is associated with *in vivo* N-terminal cleavage of CXCR4 and loss of chemotactic response to SDF-1/CXCL12 [72]. N-terminal processing of SDF-1/CXCL12 reduces receptor binding and functional activity [100] and can be achieved efficiently by PMN derived proteases including MMP-9 [74], NE [79], CG [73] and DPP-IV [83, 101] which have been implicated in hematopoietic stem and progenitor cell mobilization. In addition, CD26 (DPP-IV)-mediated processing of SDF-1 and generation of SDF-1(3–67) that is an antagonists at CXCR4 [83] suggests that N-terminal truncated SDF-1 peptides generated by proteases may also function as CXCR4 antagonists.

Concluding remarks

It is becoming evident that alterations in the interaction of HSC and marrow stroma by whatever means; CXCR4 cleavage, desensitization/antagonism, protease processing of SDF-1/CXCL12 releasing CXCR4 antagonist peptides or alteration in the SDF-1/CXCL12 gradient that retains cells in marrow, are likely responsible for stem cell mobilization. This clearly implicates the SDF-1/CXCL12/CXCR4 axis as a central component in the mobilization cascade and a role for a number of chemokines and agents that modulate this axis in the mobilization process. Their efficacy and rapid action make them attractive agents to supplement mobilization by G-CSF, particularly in allogeneic transplants that require higher stem cell yields, or in patients and normal donors who mobilize poorly to G-CSF as a single agent. This is exemplified by more robust progenitor cell mobilization by the combination of AMD plus G-CSF in Fanconi C knockout mice [88], which like Fanconi patients [20], mobilize poorly to G-CSF as a single agent. In a recent clinical trial, the combination of AMD3100 plus G-CSF was shown to be more efficacious in mobilizing sufficient cells for transplant than G-CSF alone in patients with non-Hodgkins lymphoma and multiple myeloma [102]. The number of patients reaching optimal, rather than minimal, HPC targets was increased, and fewer aphereses procedures were required to do so. In addition to combination mobilization, it is highly likely that in some cases, these SDF-1 axes modulators may also have clinical mobilizing efficacy on their own, reducing the overall time and costs associated with peripheral blood stem cell transplantation.

Lastly, the emerging role of the marrow as a reservoir for tissue specific adult stem cells or stem cells able to contribute to the repair of multiple damaged tissues raises the question of whether the chemokines that mobilize hematopoietic stem cells also have the ability to mobilize other types of stem cells as well. In this regard it has been reported that adenoviral delivered hSDF-1/CXCL12 can mobilize a bone-marrow derived circulating endothelial stem cells in mice [103]. Mobilization of endothelial stem/progenitor cells is also observed using GRO$\beta_{\Delta 4}$/CXCL2$_{\Delta 4}$ (LM

Pelus and MC Yoder, unpublished). The fact that this cell exhibits chemokinetic and chemotactic response to SDF-1/CXCL12 [103] indicates it expresses CXCR4 and therefore may be mobilized by the same mechanisms as hematopoietic stem cells, which is highly likely since endothelial and hematopoietic stem cells derive from a common cell, the hemangioblast. It remains to be determined whether chemokines or chemokine/cytokine combinations will effectively mobilize nonhematopoietic lineage stem cells.

Acknowledgements
Studies from the author's laboratories cited in this review were supported by Public Service Health Grants HL69669 and HL079654 (to LMP) and DK53674, HL67384, HL56416 and a project in PO1HL053586 (to HEB) from the National Institutes of Health.

References

1 McCredie KB, Hersch EM, Freireich EJ (1971) Cells capable of colony formation in the peripheral blood of man. *Science* 171: 293–294
2 Kurnick JE, Robinson WA (1971) Colony growth of human peripheral white blood cells *in vitro*. *Blood* 37: 136–141
3 Chervenick PA, Boggs DR (1971) *In vitro* growth of granulocytic and mononuclear cell colonies from blood of normal individuals. *Blood* 37: 131–135
4 To LB, Haylock DN, Simmons PJ, Juttner CA (1997) The biology and clinical uses of blood stem cells. *Blood* 89: 2233–2258
5 Kessinger A, Armitage JO (1991) The evolving role of autologous peripheral blood stem cell transplantation following high-dose therapy for malignancies. *Blood* 77: 211–213
6 Papayannopoulou T (2004) Current mechanistic scenarios in hematopoietic stem/progenitor cell mobilization. *Blood* 103: 1580–1585
7 Fruehauf S, Seggewiss R (2003) Its moving day: factors affecting peripheral blood stem cell mobilization and strategies for improvement. *Br J Haematol* 122: 360–375
8 Thomas J, Liu F, Link DC (2002) Mechanisms of mobilization of hematopoietic progenitors with granulocyte colony-stimulating factor. *Curr Opin Hematol* 9: 183–189
9 Goldman JM, Horowitz MM (2002) The international bone marrow registry. *Int J Hematol* 76: 393–397
10 Ringden O, Remberger M, Runde V, Bornhausser M, Blau IW, Basara N, Holig K, Beelen DW, Hagglund H, Basu O et al (2000) Faster engraftment of neutrophils and platelets with peripheral blood stem cells from unrelated donors: a comparison with marrow transplantation. *Bone Marrow Transplant* 26: 6–8
11 Benito AI, Diaz MA, Gonzales-Vicent M, Sevilla J, Madero L (2004) Hematopoietic

stem cell transplantation using umbilical cord blood progenitors: review of current clinical results. *Bone Marrow Transplant* 33: 675–690

12 Bensinger W, Singer J, Appelbaum FR, Lilleby K, Longin K, Rowley S, Clarke E, Clift R, Hansen J, Shields T et al (1993) Autologous transplantation with peripheral blood mononuclear cells collected after administration of recombinant granulocyte stimulating factor. *Blood* 81: 3158–3163

13 Blume KG, Thomas ED (2000) A review of autologous hematopoietic cell transplantation. Biol *Bone Marrow Transplant* 6: 1–12

14 Kennedy J (1998) Peripheral blood progenitor cell mobilization: A clinical review. *Pharmacotherapy* 18: 3–8

15 Welte K, Gabrilove J, Bronchud MH, Platzer E, Morstyn G (1996) Filgrastim (r-metHuG-CSF): The first 10 years. *Blood* 88: 1907–1929

16 Aversa F, Tabilio A, Velardi A, Cunningham I, Terenzi A, Falzetti F, Ruggeri L, Barbabietola G, Aristei C, Latini P et al (1998) Treatment of high risk acute leukemia with T-cell-depleted stem cells from related donors with one fully mismatched HLA haplotype. *N Eng J Med* 339: 1186–1193

17 Reisner Y, Martelli MF (2000) Transplantation tolerance induced by "mega dose" CD34+ cell transplants. *Exp Hematol* 28: 119–127

18 Bachar-Lustig E, Li HW, Gur H, Krauthgamer R, Marcus H, Reisner Y (1999) Induction of donor-type chimerism and transplantation tolerance across major histocompatibility barriers in sublethally irradiated mice by Sca-1^{+}Lin^{-} bone marrow progenitor cells: Synergisms with non-alloreactive (host x donor) F1 T cells. *Blood* 94: 3212–3221

19 Stiff P, Gingrich R, Luger S, Wyres MR, Brown RA, Lemaistre CF, Perry J, Schenkein DP, List A, Mason JR et al (2000) A randomized phase 2 study of PBPC mobilization by stem cell factor and filgrastim in heavily pretreated patients with Hodgkin's disease or non-Hodgkin's lymphoma. *Bone Marrow Transplant* 26: 471–481

20 Croop JM, Cooper R, Fernandez C, Graves V, Kreissman S, Hanenberg H, Smith FO, Williams DA (2001) Mobilization and collection of peripheral blood CD34^{+} cells from patients with Fanconi anemia. *Blood* 98: 2917–2921

21 Anderlini P, Przepiorka D, Seong C, Smith TL, Huh YO, Lauppe J, Champlin R, Kärbling M (1997) Factors affecting mobilization of CD34^{+} cells in normal donors treated with filgrastim. *Transfusion* 37: 507–512

22 Schmitz N, Dreger P, Suttorp M, Rohwedder EB, Haferlach T, Löffler H, Hunter A, Russell NH (1995) Primary transplantation of allogeneic peripheral blood progenitor cells mobilized by Filgrastim (Granulocyte Colony-Stimulating Factor). *Blood* 85: 1666–1672

23 Lataillade JJ, Clay D, Dupuy C, Rigal S, Jasmin C, Bourin P, Bousse-Kerdiles M-CL (2000) Chemokine SDF-1 enhances circulating CD34^{+} cell proliferation in synergy with cytokines: possible role in progenitor survival. *Blood* 95: 756–768

24 Broxmeyer HE, Kohli L, Kim CH, Lee Y, Mantel C, Cooper S, Hangoc G, Shaheen M, Li X, Clapp DW (2003) Stromal cell derived factor-1/CXCL12 enhances survival/anti-

apoptosis of hematopoietic stem and progenitor cells: direct effects mediated through CXCR4 and Gαi proteins. *J Leukocyte Biol* 73: 630–638

25 Broxmeyer HE, Cooper S, Kohli L, Hangoc G, Lee YH, Mantel C, Clapp DW, Kim CH (2003) Transgenic expression of stromal cell derived factor-1/CXCL12 enhances myeloid progenitor cell survival/anti-apoptosis *in vitro* in response to growth factor withdrawal and enhances myelopoiesis. *J Immunol* 170: 421–429

26 Lee Y, Gotoh, Kwon HJ, You M, Kohli L, Mantel C, Cooper S, Hangoc G, Miyazawa K, Ohyashiki K, Broxmeyer HE (2002) Enhancement of intracellular signaling associated with hematopoietic progenitor cell survival in response to SDF-1/CXCL12 in synergy with other cytokines. *Blood* 99: 4307–4317

27 Lataillade JJ, Clay D, Bourin P, Herodin F, Dupuy C, Jasmin C, Le Bousse-Kerdiles M-C (2002) Stromal cell-derived factor 1 regulates primitive hematopoiesis by suppressing apoptosis and by promoting G0/G1 transition in CD34$^+$ cells: evidence for an autocrine/paracrine mechanism. *Blood* 99: 1117–1129

28 Broxmeyer HE, Cooper S, Hangoc G, Kim KC (2005) Stromal cell-derived factor-1/CXCL12 selectively counteracts inhibitory effects of myelosuppressive chemokines on hematopoietic progenitor cell proliferation *in vitro*. *Stem Cells Dev* 14: 199–203

29 Aiuti A, Webb IJ, Bleul C, Springer T, Gutierrez-Ramos JC (1997) The chemokine SDF-1 is a chemoattractant for human CD34$^+$ hematopoietic progenitor cells and provides a new mechanism to explain the mobilization of CD34$^+$ progenitors to peripheral blood. *J Exp Med* 185: 111–120

30 Kim CH, Broxmeyer HE (1998) *In vitro* behavior of hematopoietic progenitor cells under the influence of chemoattractants: Stromal cell-derived factor-1, steel factor, and the bone marrow environment. *Blood* 91: 100–110

31 Möhle R, Bautz F, Rafi S, Moore MAS, Brugger W, Kanz L (1998) The chemokine receptor CXCR4 is expressed on CD34$^+$ hematopoietic progenitors and leukemic cells and mediates transendothelial migration induced by stromal cell-derived factor-1. *Blood* 91: 4523–4530

32 Peled A, Petit I, Kollet O, Magid M, Ponomaryov T, Byk T, Nagler A, Ben-Hur H, Many A, Shultz L et al (1999) Dependence of human stem cell engraftment and repopulation of NOD/SCID mice on CXCR4. *Science* 283: 845–848

33 Lapidot T, Kollet O (2002) The essential roles of the chemokine SDF-1 and its receptor CXCR4 in human stem cell homing and repopulation of transplanted immune-deficient NOD/SCID and NOD/SCID/β2m^{null} mice. *Leukemia* 16: 1992–2203

34 Christopherson KW II, Hangoc G, Mantel CR, Broxmeyer HE (2004) Modulation of hematopoietic stem cell homing and engraftment by CD26. *Science* 305: 1000–1003

35 Guo Y, Hangoc G, Bian H, Pelus LM, Broxmeyer HE (2005) SDF-1/CXCL12 enhances survival and chemotaxis of murine embryonic stem cells and production of primitive and definitive hematopoietic progenitor cells. *Stem Cells* 23: 1324–1332

36 Lord BI, Woolford LB, Wood LM, Czaplewski LG, McCourt M, Hunter MG, Edwards RM (1995) Mobilization of early hematopoietic progenitor cells with BB-10010: A

genetically engineered variant of human macrophage inflammatory protein-1α. *Blood* 85: 3412–3415

37 Broxmeyer HE, Orazi A, Hague NL, Sledge GW Jr., Rasmussen H, Gordon MS (1998) Myeloid progenitor cell proliferation and mobilization effects of BB10010, a genetically engineered variant of human macrophage inflammatory protein-1α, in a phase I clinical trial in patients with relapsed/refractory breast cancer. *Blood Cells, Molec, and Dis* 24: 14–30

38 Broxmeyer HE, Cooper S, Hangoc G, Gao J-L, Murphy PM (1999) Dominant myelopoietic effector functions mediated by chemokine receptor CCR1. *J Exp Med* 189: 1987–1992

39 Laterveer L, Lindley IJD, Hamilton MS, Willemze R, Fibbe WE (1995) Interleukin-8 induces rapid mobilization of hematopoietic stem cells with radioprotective capacity and long-term myelolymphoid repopulating ability. *Blood* 85: 2269–2275

40 Fibbe WE, Pruijt JFM, Velders G, Opdenakker G, van Kooyk Y, Figdor CG, Willemze R (1999) Biology of IL-8-induced stem cell mobilization. *Annals NY Acad Sci* 872: 71–82

41 Laterveer L, Zijlmans JMJM, Lindley IJD, Hamilton MS, Willemze R, Fibbe WE (1996) Improved survival of lethally irradiated recipient mice transplanted with circulating progenitor cells mobilized by IL-8 after pretreatment with stem cell factor. *Exp Hematol* 24: 1387–1393

42 King AG, Horowitz D, Levin SB, Farese AM, MacVittie TJ, Pelus LM (2001) Rapid mobilization of murine hematopoietic stem cells with enhanced engraftment properties and evaluation of hematopoietic progenitor cell mobilization in rhesus monkeys by a single injection of SB-251353, a specific truncated form of the human CXC chemokine GROβ. *Blood* 97: 1534–1542

43 Pelus LM, Horowitz D, Cooper SC, King AG (2002) Peripheral blood stem cell mobilization. A role for CXC chemokines. *Crit Rev Oncol Hematol* 43: 257–275

44 King AG, Johanson K, Frey CA, DeMarsh PL, White JR, McDevitt P, McNulty D, Balcarek J, Jonak ZL, Bhatnagar PK et al (2000) Identification of unique truncated KC/GROβ chemokines with potent hematopoietic and anti-infective activities. *J Immunol* 164: 3774–3782

45 Pelus LM, Bian H, King AG, Fukuda S (2004) Neutrophil-derived MMP-9 mediates synergistic mobilization of hematopoietic stem and progenitor cells by the combination of G-CSF and the chemokines GROβ/CXCL2 and GROβT/CXCL2$_{\Delta 4}$. *Blood* 103: 110–119

46 Wang J-B, Mukaida N, Zhang Y, Ito T, Nakao S, Matsushima K (1997) Enhanced mobilization of hematopoietic progenitor cells by mouse MIP-2 and granulocyte colony-stimulating factor in mice. *J Leukocyte Biol* 62: 503–509

47 Fibbe WE, Pruijt JFM, Kooyk YV, Figdor CG, Opdenakker G, Willemze R (2000) The role of metalloproteinases and adhesion molecules in interleukin-8-induced stem cell mobilization. *Sem Hematol* 37: 19–24

48 Broxmeyer HE, Cooper S, Cacalano G, Hague NL, Bailish E, Moore MW (1996)

Involvement of Interleukin (IL) 8 receptor in negative regulation of myeloid progenitor cells *in vivo*: evidence from mice lacking the murine IL-8 receptor homologue. *J Exp Med* 184: 1825–1832

49 Laterveer L, Lindley IJD, Heemskerk DPM, Camps JAJ, Pauwels EKJ, Willemze R, Fibbe WE (1996) Rapid mobilization of hematopoietic progenitor cells in rhesus monkeys by a single intravenous injection of Interleukin-8. *Blood* 87: 781–788

50 Pruijt JFM, Verzaal P, van Os R, de Kruijf E-JFM, van Schie MLJ, Mantovani A, Vecchi A, Lindley IJD, Willemze R, Starchx S, Opendaker G, Fibbe WE (2002) Neutrophils are indispensable for hematopoietic stem cell mobilization induced by interleukin-8 in mice. *Proc Natl Acad Sci USA* 99: 6228–6233

51 Pruijt JFM, van Kooyk Y, Figdor CG, Lindley IJD, Willemze R, Fibbe WE (1998) Anti-LFA-1 blocking antibodies prevent mobilization of hematopoietic progenitor cells induced by interleukin-8. *Blood* 91: 4099–4105

52 Pruijt JFM, Fibbe WE, Laterveer L, Pieters RA, Lindley IJD, Paemen L, Masure S, Willemze R, Opdenakker G (1999) Prevention of interleukin-8-induced mobilization of hematopoietic progenitor cells in rhesus monkeys by inhibitory antibodies against the Metalloproteinase gelatinase B (MMP-9). *Proc Natl Acad Sci USA* 96: 10863–10868

53 Anderson DC, Miller LJ, Schmalstieg FC, Rothlein R, Springer TA (1986) Contributions of the Mac-1glycoprotein family to adherence-dependent granulocyte functions: structure-function assessments employing subunit-specific monoclonal antibodies. *J Immunol* 137: 15–27

54 Borregaard N, Cowland JB (2000) Granules of the human neutrophilic polymorphonuclear leukocyte. *Blood* 89: 3503–3521

55 Papayannopoulou T, Priestly GV, Nakamoto B, Zafiropoulos V, Scott LM, Harlan JM (2001) Synergistic mobilization of hematopoietic progenitor cells using concurrent β1 and β2 integrin blockade or β2-deficient mice. *Blood* 97: 1282–1288

56 Velders GA, Pruijt JFM, Verzaal P, van Os R, van Kooyk Y, Figdor CG, de Kruijf E-JFM, Willemze R, Fibbe WE (2002) Enhancement of G-CSF-induced stem cell mobilization by antibodies against the β2 integrins LFA-1 and Mac-1. *Blood* 327: 327–333

57 Werb Z (1997) ECM and cell surface proteolysis: Regulating cellular ecology. *Cell* 91: 439–442

58 Birkedal-Hansen H, Moore WG, Bodden MK, Windsor LJ, Birkedal-Hansen B, DeCarlo A, Engler JA (1993) Matrix metalloproteinases: a review. *Crit Rev Oral Biol Med* 4: 197–250

59 Nelson AR, Fingleton B, Rothenberg ML, Matrisian LM (2000) Matrix metalloproteinases: Biological activity and clinical implications. *J Clin Oncol* 18: 1135–1149

60 Ramos-Desimone N, Hahn-Dantona E, Sipley J, Nagase H, French DL, Quigley JP (1999) Activation of matrix metalloproteinase-9 (MMP-9) via a converging plasmin/stromelysin-1 cascade enhances tumor cell invasion. *J Biol Chem* 274: 13066–13076

61 Fukuda S, Bian H, King AG, Pelus LM (2005) The CXC chemokine GROβ mobilizes

PBSC with enhanced homing capacity independent of the SDF-1/CXCR4 axis. *Submitted*

62 Kollet O, Spiegel A, Peled A, Petit I, Byk T, Hershkoviz R, Guetta E, Barkai G, Nagler A, Lapidot T (2001) Rapid and efficient homing of human CD34(+)CD38(–/low) CXCR4(+) stem and progenitor cells to the bone marrow and spleen of NOD/SCID and NOD/SCID/β2m(null) mice. *Blood* 97: 3283–3291

63 Shen H, Cheng T, Olszak I, Garcia-Zepeda E, Lu Z, Hermann S, Fallon R, Luster AD, Scadden DT (2001) CXCR-4 desensitization is associated with tissue localization of hematopoietic progenitor cells. *J Immunol* 166: 5027–5033

64 Hattori K, Heissig B, Tashiro T, Tateno M, Shieh J-H, Hackett NR, Quitoriano MS, Crystal RG, Rafii S, Moore MAS (2001) Plasma elevation of stromal cell-derived factor-1 induces mobilization of mature and immature hematopoietic progenitor and stem cells. *Blood* 97: 3354–3360

65 Ma Q, Jones D, Springer TA (1999) The chemokine receptor CXCR4 is required for the retention of B lineage and granulocytic precursors within the bone marrow microenvironment. *Immunity* 10: 463–471

66 Nagasawa T, Hirota S, Tachibaba K, Takakura N, Nishikawa S, Kitamura Y, Yoshida N, Kikutani H, Kishimoto T (1996) Defects of B-cell lymphopoiesis and bone-marrow myelopoiesis in mice lacking the CXC chemokine PBSF/SDF-1. *Nature* 382: 635–638

67 Zou Y-R, Kottmann AH, Kuroda M, Taniuchi I, Littman DR (1998) Function of the chemokine receptor CXCR4 in haematopoiesis and in cerebellar development. *Nature* 393: 595–599

68 Peled A, Grabovsky V, Habler L, Sandbank J, Arenzana-Seisdedos F, Petit I, Ben-Hur H, Lapidot T, Alon R (1999) The chemokine SDF-1 stimulates integrin-mediated arrest of CD34$^+$ cells on vascular endothelium under shear flow. *J Clin Invest* 104: 1199–1211

69 Hidalgo A, Sanz-Rodríguez F, Rodríguez-Fernández JL, Albella B, Blaya C, Wright N, Cabañas C, Prósper F, Gutierrez-Ramos JC, Teixidó J (2001) Chemokine stromal cell-derived factor-1α modulates VLA-4 integrin-dependent adhesion to fibronectin and VCAM-1 on bone marrow hematopoietic progenitor cells. *Exp Hematol* 29: 345–355

70 Shen W, Bendall LJ, Gottlieb DJ, Bradstock KF (2001) The chemokine receptor CXCR4 enhances integrin-mediated *in vitro* adhesion and facilitates engraftment of leukemic precursor-B cells in the bone marrow. *Exp Hematol* 29: 1439–1447

71 Petit I, Szyper-Kravitz M, Nagler A, Lahav M, Peled A, Habler L, Ponomaryov T, Taichman RS, Arenzana-Seisdedos F, Fujii N et al (2002) G-CSF induces stem cell mobilization by decreasing bone marrow SDF-1 and up-regulating CXCR4. *Nature Immunology* 3: 687–694

72 Lévesque J-P, Hendy J, Takamatsu Y, Simmons PJ, Bendall LJ (2003) Disruption of the CXCR4/CXCL12 chemotactic interaction during hematopoietic stem cell mobilization induced by G-CSF or cyclophosphamide. *J Clin Invest* 111: 187–196

73 Delgado MB, Clark-Lewis I, Loescher P, Langen H, Thelen M, Baggiolini M, Wolf M (2001) Rapid inactivation of stromal cell-derived factor-1 by cathepsin G associated with lymphocytes. *Eur J Immunol* 31: 699–707

74 McQuibban GA, Butler GS, Gong J-H, Bendall L, Powers C, Clark-Lewis I, Ovrall CM (2001) Matrix metalloproteinase activity inactivates the CXC chemokine stromal cell-derived factor-1. *J Cell Biol* 276: 43503–43508

75 Semerad CL, Liu F, Gregory AD, Stumpf K, Link DC (2002) G-CSF is an essential regulator of neutrophil trafficking from the bone marrow to the blood. *Immunity* 17: 413–423

76 Semerad CL, Liu F, Christopher MJ, Link DC (2003) G-CSF treatment induces a significant decrease in CXCL12 mRNA expression in the bone marrow. *Blood* 102: 824a

77 Christopher MJ, Liu F, Link DC (2004) Disruption of SDF-1/CXCR4 signaling during Flt-3 ligand and stem cell factor (SCF) induced hematopoietic progenitor mobilization. *Blood* 104: 121b

78 Pelus LM, Bian H, Fukuda S, Wong D, Merzouk A, Salari H (2005) The CXCR4 agonist peptide, CTCE-0021, rapidly mobilizes polymorphonuclear neutrophils and hematopoietic progenitor cells into peripheral blood and synergizes with granulocyte colony-stimulating factor. *Exp Hematol* 33: 295–307

79 Valenzuela-Fernández A, Planchenault T, Baleux F, Staropoli I, LeBarillec K, Leduc D, Deluanay T, Lazarini F, Virelizier J-L, Chignard M et al (2002) Leukocyte elastase negatively regulates stromal cell-derived factor-1 (SDF-1)/CXCR4 binding and functions by amino-terminal processing of SDF-1 and CXCR4. *J Biol Chem* 277: 15677–15689

80 Merzouk A, Wong D, Salari H, Bian H, Fukuda S, Pelus LM (2004) Rational design of chemokine SDF-1 analogs with agonist activity for the CXCR4 receptor and the capacity to rapidly mobilize PMN and hematopoietic progenitor cells in mice. *Lett Drug Des Discov* 1: 126–134

81 Zhong R, Law P, Wong D, Merzouk A, Salari H, Ball ED (2004) Small peptide analogs to stromal derived factor-1 enhance chemotactic migration of human and mouse hematopoietic cells. *Exp Hematol* 32: 470–475

82 Christopherson K II, Cooper S, Broxmeyer HE (2003) Cell surface peptidase CD26/DPPIV mediates G-CSF mobilization of mouse progenitor cells. *Blood* 101: 4680–4686

83 Christopherson K II, Hangoc G, Broxmeyer HE (2002) Cell surface peptidase CD26/Dipeptidylppeptidase IV regulates CXCL12/Stromal cell-derived factor-1α-mediated chemotaxis of human cord blood $CD34^+$ progenitor cells. *J Immunol* 169: 7000–7008

84 Christopherson KW II, Cooper S, Broxmeyer HE (2003) CD26 is essential for normal G-CSF-induced progenitor cell mobilization as determined by $CD26^{-/-}$ mice. *Exp Hematol* 31: 1126–1134

85 De Clercq E (2003) The bicyclan AMD3100 story. *Nat Rev Drug Discov* 2: 581–587

86 Hatse S, Princen K, Bridger G, De Clercq E, Schols D (2002) Chemokine receptor inhibition by AMD3100 is strictly confined to CXCR4. *FEBS Letters* 527: 255–262

87 Joo EK, Broxmeyer HE, Kwon HJ, Kang HB, Kim JS, Lim JS, Choe YK, Choe IS, Myung PK, Lee Y (2004) Enhancement of cell survival by stromal cell-derived factor-

1/CXCL12 involves activation of CREB and induction of Mcl-1 and c-Fos in factor-dependent human cell line MO7e. *Stem Cells Dev* 13: 563–570
88 Broxmeyer HE, Clapp DW, Orschell CM, Hangoc G, Cooper S, Plett A, Liles WC, Li X, Graham-Evans B, Calandra G et al (2005) Rapid mobilization of murine and human hematopoietic stem and progenitor cells with AMD3100, a CXCR4 antagonist. *J Exp Med* 201: 1307–1318
89 Liles WC, Broxmeyer HE, Rodger E, Wood B, Hübel K, Cooper S, Hangoc G, Bridger GJ, Henson GW, Calandra G et al (2003) Mobilization of hematopoietic progenitor cells in healthy volunteers by AMD3100, a CXCR4 antagonist. *Blood* 102: 2728–2730
90 Hubel K, Liles WC, Broxmeyer HE, Rodger E, Wood B, Cooper S, Hangoc G, MacFarland R, Bridger GJ, Henson GW et al (2004) Leukocytosis and mobilization of $CD34^+$ hematopoietic progenitor cells by AMD3100, a CXCR4 antagonist. *Supportive Cancer Therapy* 1: 165–172
91 Devine SM, Flomenberg N, Vesole DH, Liesveld J, Weisdoef D, Badel K, Calandra G, Dipersio JF (2004) Rapid mobilization of $CD34^+$ cells following administration of the CXCR4 antagonist AMD3100 to patients with multiple myeloma and Non-Hodgkin's lymphoma. *J Clin Oncol* 22: 1095–1102
92 Liles WC, Rodger E, Broxmeyer HE, Dehner C, Badel K, Calandra J, Christensen J, Wood B, Price TH, Dale DC (2004) Augmented mobilization and collection of CD34+ hematopoietic cells from normal human volunteers stimulated with G-CSF by single-dose administration of AMD3100, a CXCR4 antagonist. *Transfusion* 45: 295–300
93 Yang OO, Swanberg SL, Lu Z, Dziejman M, McCoy J, Luster AD, Walker BD, Herrmann SH (1999) Enhanced inhibition of human immunodeficiency virus type 1 by met-stromal derived factor 1β correlates with down-modulation of CXCR4. *J Virology* 73: 4582–4589
94 Basu S, Broxmeyer HE (2005) Transforming growth factor-{beta}1 modulates responses of $CD34^+$ cord blood cells to stromal cell-derived factor-1/CXCL12. *Blood* 106: 485–493
95 Murdoch C, Finn A (2000) Chemokine receptors and their role in inflammation and infectious diseases. *Blood* 95: 3032–3043
96 Liu F, Wu HY, Wesselschmidt RKT, Link DC (1996) Impaired production and increased apoptosis of neutrophils in granulocyte colony-stimulating factor receptor-deficient mice. *Immunity* 5: 491–501
97 Lévesque J-P, Liu F, Simmons PJ, Betsuyaku T, Senior RM, Pham C, Link DC (2004) Characterization of hematopoietic progenitor mobilization in protease-deficient mice. *Blood* 104: 65–72
98 Lévesque J-P, Takamatsu Y, Nilsson SK, Haylock DN, Simmons PJ (2001) Vascular cell adhesion molecule-1 (CD106) is cleaved by neutrophil proteases in the bone marrow following hematopoietic progenitor cell mobilization by granulocyte colony-stimulating factor. *Blood* 98: 1289–1297
99 Lévesque J-P, Hendy J, Takamatsu Y, Williams B, Winkler IG, Simmons PJ (2002) Mobilization by either cyclophosphamide or granulocyte colony-stimulating factor trans-

forms the bone marrow into a highly proteolytic environment. *Exp Hematol* 30: 440–449

100 Baggiolini M (1998) Chemokines and leukocyte traffic. *Nature* 392: 565–568

101 Proost P, Struyf S, Schols D, Durinx C, Wuyts A, Lenaerts J-P, DeClercq E, DeMeester I, VanDamme J (1998) Processing by CD26/dipeptidyl-peptidase IV reduces the chemotactic and anti-HIV-1 activity of stromal-cell derived factor-1α. *FEBS Letters* 432: 73–76

102 Flomenberg N, Devine SM, DiPersio JF, Liesveld JL, McCarty JM, Rowley SD, Vesole DH, Badel K, Calandra G (2005) The use of AMD3100 plus G-CSF for autologous hematopoietic progenitor cell mobilization is superior to G-CSF alone. *Blood* 106: 1867–1874

103 Moore MAS, Hattori K, Heissig B, Shieh J-H, Dias S, Crystal RG, Rafii S (2001) Mobilization of endothelial and hematopoietic stem and progenitor cells by adenovector-mediated elevation of serum levels of SDF-1, VEGF and angiopoietin-1. *Annals NY Acad Sci* 938: 36–45

CCR5 antagonists: from discovery to clinical efficacy

Shon R. Pulley

Eli Lilly and Company, Lilly Corporate Center, Discovery Chemistry & Research Technologies, Indianapolis, IN 46285, USA

Introduction

CCR5 is a prototypically inflammatory chemokine receptor belonging to the seven transmembrane G-protein-coupled receptor (GPCR) family. This family is generally considered druggable and well represented in marketed drugs [1–6]. The CCR5 receptor is expressed on numerous host defense cells including monocytes, macrophages, T-lymphocytes, dendritic cells and microglia [4, 5]. Interaction of CCR5 with its ligands MIP-1α, MIP-1β (CCL3/CCL4) or RANTES (CCL5) results in a conformational change in the seven transmembrane domain initiating a signaling cascade through heterotrimeric G-proteins ultimately giving rise to migration of immune cells to sites of inflammation [2, 5]. Due to the well documented role of CCR5 in the immune system, it has been implicated in the pathophysiology of rheumatoid arthritis (RA), multiple sclerosis (MS), transplant rejection, gastric disorders, diabetes and myeloma [7]. Perhaps the biggest driver behind the development of CCR5 receptor antagonists was the discovery that CCR5 plays an important role as a co-receptor for macrophage tropic HIV-1 strains to facilitate viral fusion and entry into host cells [5, 8–12]. In addition, individuals with a mutation in the CCR5 gene (CCR5Δ32) lacking expression of CCR5 on the cell surface are resistance to HIV-1 infection without notable immune system effects [13, 14], thus validating pharmaceutical intervention with a CCR5 antagonist.

Given the presence of CCR5 at the cell surface, both large molecule (protein/peptide) and small molecule (agonist/antagonist) strategies have the potential to manipulate the CCR5/chemokine axis. Large molecule CCR5 ligands include amino-terminus modified RANTES derivatives, such as (PSC)-RANTES, and AOP-RANTES have been shown to block HIV infection through a process generally thought to induce receptor internalization [15, 16]. Monoclonal antibody antagonists Pro-140 (Progenics) and HGS-Ab004 (Human Genome Science) are both reported in Phase 1 clinical trials and block HIV-1 entry by binding to the second extracellular loop of CCR5. In the case of Pro-140 the inhibition of HIV entry is maintained without

Chemokine Biology – Basic Research and Clinical Application, Volume II
edited by Kuldeep Neote, Gordon L. Letts and Bernhard Moser

compromising the function of the CCR5 receptor [17, 18]. Small molecule antagonists of CCR5 are thought to interact with the receptor in an allosteric manner locking the receptor in a conformation that prohibits its co-receptor function. They also limit the signals induced by various chemokines [19]. Small molecule CCR5 antagonists Aplaviroc, Vicriviroc, and Maraviroc have advanced to human efficacy clinical trials for HIV infection in AIDS patients. The following account will summarize preclinical and clinical efforts toward the development of small molecule antagonists of CCR5.

CCR5 antagonists prior to human efficacy studies

The journey to a clinical application of a CCR5 antagonist began with the report that CCR5 plays a critical role in the early phase of HIV infection [5, 8, 9, 11, 12]. This triggered efforts by Pfizer, Schering-Plough, and GSK/Ono that have led to small molecule CCR5 antagonists that are in human efficacy trials. Efforts by Takeda, Novartis, AztraZeneca, Merck, Incyte and others have varying degrees of clinical and preclinical information available. AnorMed has a substantial presence in the anti-HIV area with the CXCR4 inhibitor AMD-070 currently in Phase II clinical trials for HIV and have disclosed an active CCR5 program [20].

Prior to the identification of CCR5's role in HIV infection and other disease states, many pharmaceutical companies built substantial collections of compounds targeting GPCRs. These compounds offered the potential of providing quality starting points for CCR5 antagonists. Screening and/or rational design by the industry led to the identification of a number of initial leads some of which are represented in Figure 1 [21]. Many of these initial leads share a notable feature – a positively charged nitrogen found in a so called 'privileged' structure common in small molecule inhibitors of GPCRs [22]. These initial compounds required further optimization for potency, pharmacokinetic properties and selectivity *versus* other GPCR receptors.

AstraZeneca reported 1-(3,3-diphenylpropyl)-piperidine **1** (Fig. 1) as weak inhibitor of CCR5. They demonstrated that the 2,3-dihydroisoindol-1-one was not necessary to maintain potency and further optimization resulted in compound 7 (RANTES IC_{50} = 37 nM, Fig. 2). Compound 7 and close analogs demonstrated human muscarinic, serotonergic and *h*ERG activity suggesting further optimization for selectivity was necessary [23]. In an effort to further improve potency, SAR optimization of the 3,3-diphenyl propyl region resulted in compound 8 with good affinity (IC_{50} = 1.7 nM) and with modest oral exposure [24]. No further improvements for this series are reported.

Researchers at Novartis reasoned that the thienylmethyl phenylpiperidine of compound 3 (Fig. 1) served as a replacement for the phenylpiperidine ethyl oxime found in Schering-Plough's SCH-C *vide infra* [25]. The introduction of a methyl

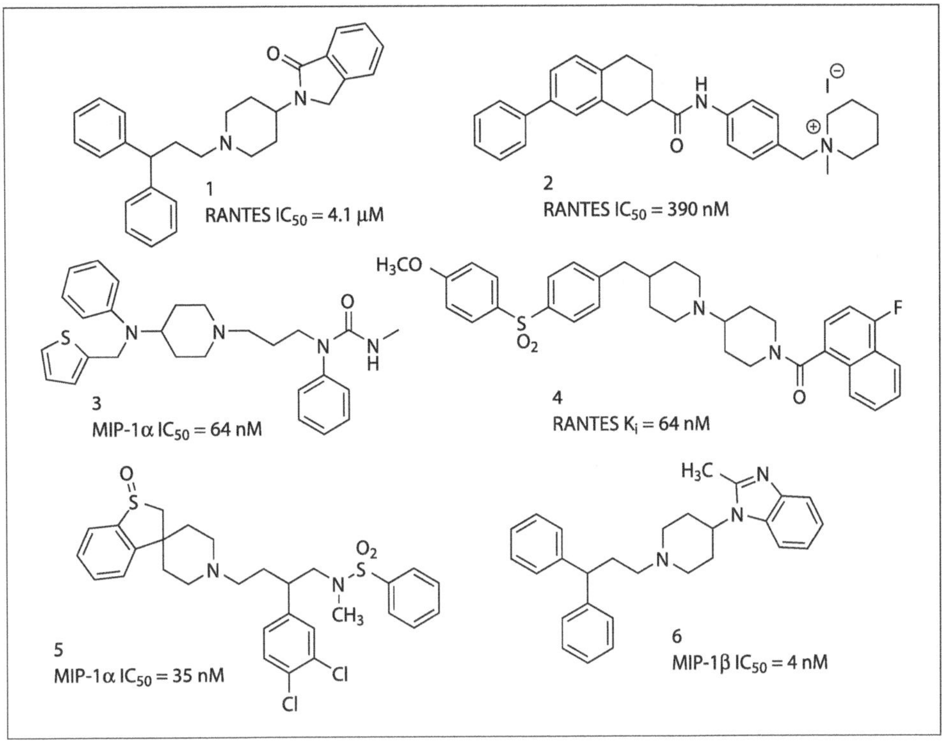

Figure 1

group and a second piperidine followed by subsequent optimization of the N-terminal substitution ultimately resulted in compound 9. Compound 9 has low nanomolar CCR5 inhibitory activity in both human and cynomologous monkeys (MIP-1α IC_{50} = 2.9 and 12 nM respectively, Fig. 2). Furthermore, compound 9 showed no activity up to 1 μM on a panel of chemokine receptors or other GPCRs. Additionally, compound 9 demonstrated > 50% oral bioavailability in both rats and primates suggesting a reasonable candidate for further *in vivo* profiling. However, no further development is noted.

Researchers at Merck have described their efforts toward CCR5 antagonists using compound 5 (Fig. 1) as the starting point [26]. While compound 5 had reasonable affinity in MIP-1α binding assays (IC_{50} = 35 nM), the antiviral activity was weak (HIV PBMC IC_{95} = 6–12 μM) and required optimization. SAR efforts determined the sulfonamide was necessary for high affinity and installation of a benzyl carbamoyl group at the 4-position of the piperidine yielded compound 10 demonstrating good potency in binding (MIP-1α IC_{50} = 0.1 nM) and antiviral assays (IC_{95} = 8 nM) with improved rat oral bioavailability (29%) over the original lead com-

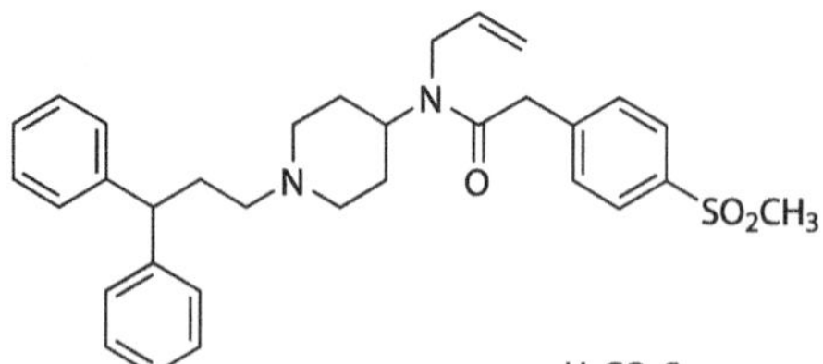

7 RANTES IC_{50} = 37 nM

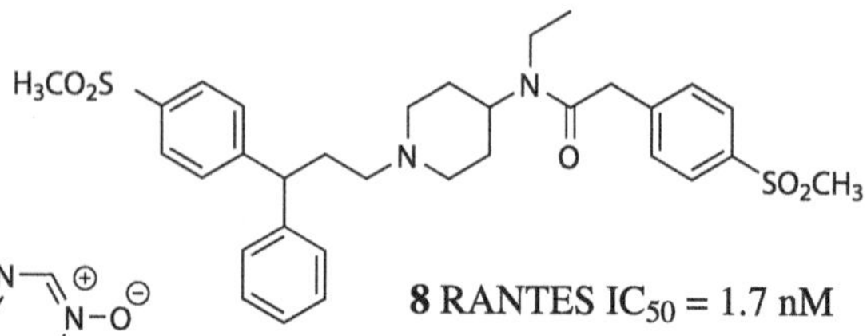

8 RANTES IC_{50} = 1.7 nM

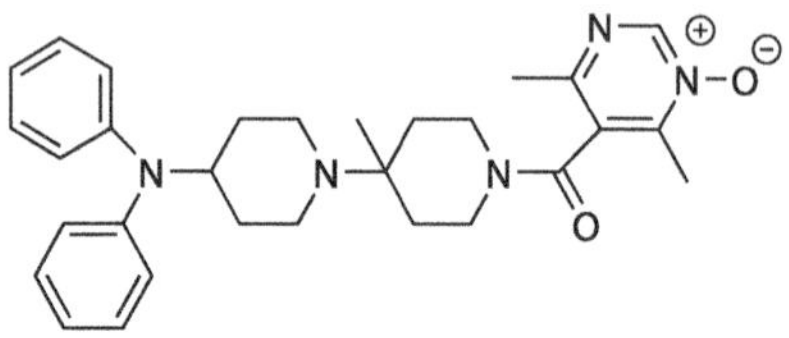

9 MIP-1α IC_{50} = 2.9 nM

Figure 2

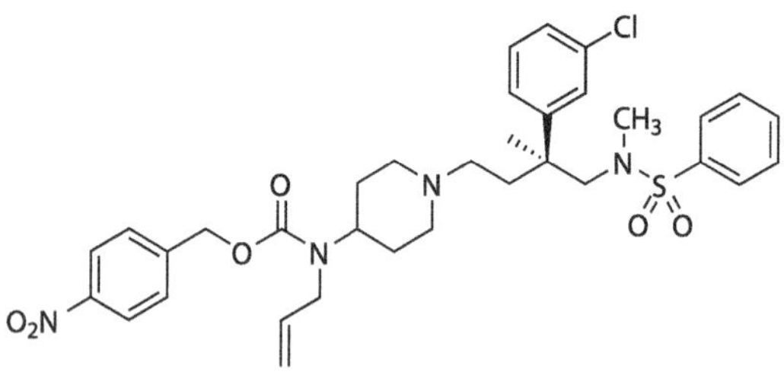

10 MIP-1α IC_{50} = 0.1nM
HIV (Bal)-PBMC: IC_{95} = 13nM

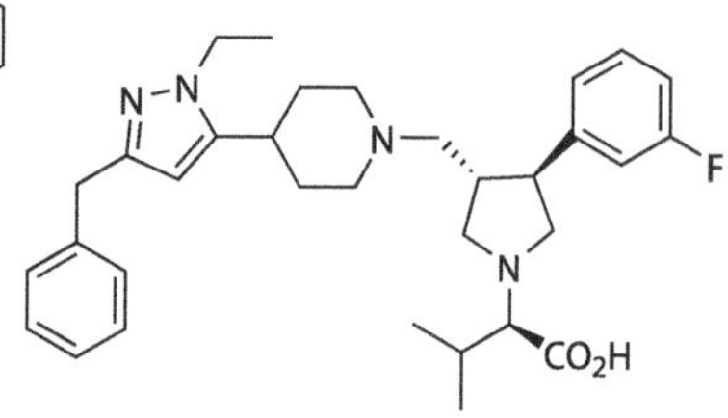

11 MIP-1α IC_{50} = 1.5 nM
HIV (Bal)-PBMC: IC_{95} = 22 nM

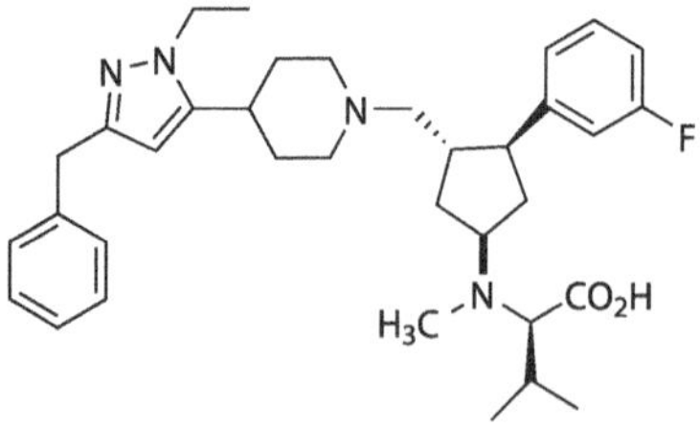

12 MIP-1α IC_{50} = 1.1 nM
HIV (Bal)-PBMC: IC_{95} = 8 nM

Figure 3

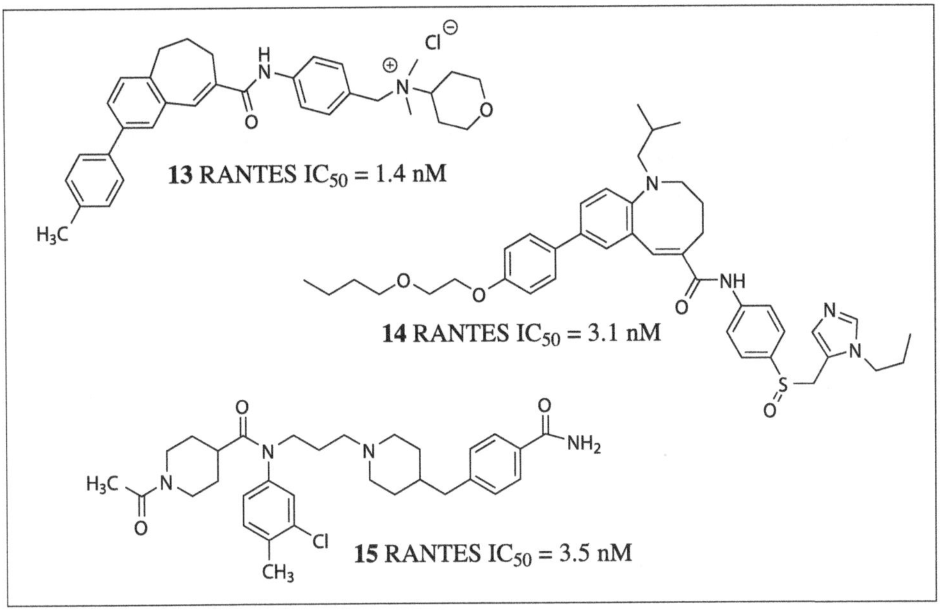

Figure 4

pound **5** (Fig. 3) [27]. Compound **10** had an excellent selectivity profile against chemokine receptors and other GPCRs, however the oral bioavailability in dog was limited (< 1%). Analysis of the active conformation of compound **10** suggested linking the *N*-methyl group to the backbone butyl chain. This resulted in a series of conformationally locked 1,3,4-trisubstituted pyrrolidines represented by compound **11** [28, 29]. Further SAR extended the pyrrolidines to amino acid substituted cyclopentanes giving rise to compound **12** (MRK-167, MIP-1α IC_{50} = 1.1 nM) which shows significant antiviral activity (HIV-Bal-PBMC: IC_{95} < 8 nM) [27, 30]. The oral bioavailability of compound **12** is reported between 43–66% in rats, dogs, and monkeys with increased exposure upon co-administration with ritonavir to mitigate CYP3A4 metabolism and P-gp intestinal efflux. It appears compound **12** is no longer being pursed by Merck as an oral anti-HIV agent and they have licensed compound **12** to The International Partnership for Microbicides (IPM) to investigate the potential of compound **12** as a topical intravaginal gel to protect women from HIV [31]. Recent meeting abstracts also suggest further work in allograft rejection therapy [32].

Takeda Pharmaceutical published the first account of small molecule CCR5 antagonists with the development of the initial screening hit compound 2 (Fig. 1) to compound **13** (TAK-779) [33]. Compound **13** (TAK-779) was potent in both binding (RANTES IC_{50} = 1.4 nM) and *in vitro* antiviral assays (Fig. 4). The poor

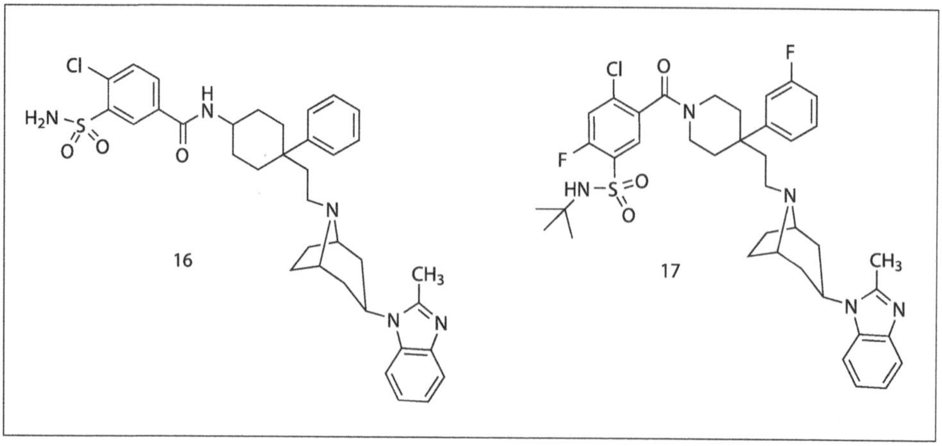

Figure 5

pharmacokinetics of the quaternary ammonium moiety led to initial clinical work using a subcutaneous injection. Further development was halted when injection site reactions were observed. In an effort to improve pharmacokinetic properties, the quaternary ammonium side chain was replaced with a sulfoxide moiety, followed by increasing the fused ring size and replacing the aryl methyl in compound **13** to give compound **14** (TAK-652). Compound **14** is active against a number of R5-HIV isolates (EC_{90} = 0.25 nM) and well tolerated in healthy volunteers up to 100 mg [34]. However compound **14** demonstrated MCP-1 inhibition at CCR2 and thus is not truly CCR5 selective. Takeda also disclosed compound **15** (TAK-220) in their efforts toward a clinical application of a CCR5 antagonist [35, 36]. Compound **15** was evaluated in combination with anti-retrovirals *in vitro* and a favorable synergy was observed suggesting further evaluation [37]. No human efficacy results with compound **14** (TAK-652) or compound **15** (TAK-220) are thus far reported.

Prior to partnering with Ono Pharmaceuticals in the development of Aplaviroc *vide infra*, researchers at GSK disclosed a number of compounds originating from compound **16** (Fig. 5) [9, 38]. Compound **16** demonstrated activity in HIV-1 Bal/HOS-CD4 cellular assays (IC_{50} = 78 nM). Replacing the aminocyclohexane with a 4,4'-disubstituted piperidine and further SAR optimization resulted in compound **17** (GSK-929, IC_{50} = 3.8 nM). While little information beyond patent literature is available for these compounds, recent meeting abstracts suggest considerable optimization was done around pharmacokinetics, *h*ERG blockade and antiviral properties [39].

In a recent announcement, Incyte has started Phase 1 trails with INCB-9471 [40]. The double-blinded study is structured to evaluate the safety and pharmaco-

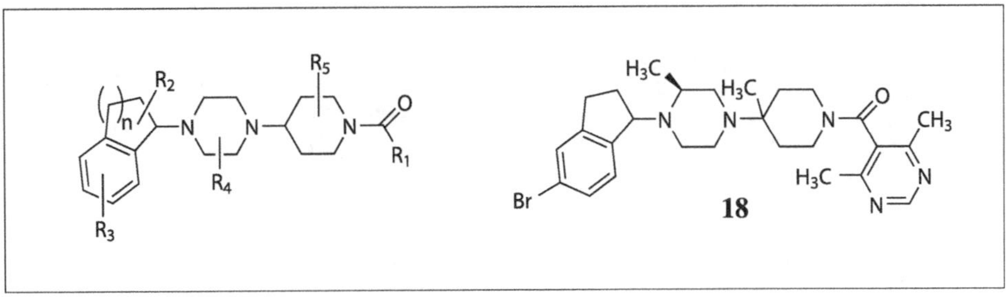

Figure 6

kinetic profile of INCB-9471 in healthy volunteers. The structure of INCB-9471 has not been disclosed but Incyte recently disclosed piperizinyl piperidines in WO 2005/101838 represented by the generic structure and compound **18** (Fig. 6). INCB-9471 is reported to inhibit ligand binding to CCR5 and R5-HIV infection at concentrations < 10 nM. INCB-9471 is reported to demonstrate selectivity against other chemokines, GPCRs, ion channels, enzymes, and transporters. Oral pharmacokinetics in rat, dog, and monkeys suggest, low clearance, moderate volumes, and excellent bioavailability > 95%. Phase IIa studies with treatment-naïve and minimally treatment experienced HIV-infected subjects are expected to begin in the forth quarter of 2006 [41].

CCR5 antagonists reaching human efficacy studies

The race for an approved small molecule CCR5 antagonist has involved three competitors, Pfizer's Maraviroc, Schering-Plough's Vicriviroc and GSK's Aplaviroc. All three compounds began human efficacy trials and at the time of this account, only Maraviroc has received a recommendation by the Drug Safety Monitoring Board (DSMB) to continue all Phase III trails as originally designed. Vicriviroc and Aplaviroc have encountered efficacy and/or toxicity issues that have limited or halted further development of these compounds. Reviews covering the development of the latter CCR5 antagonists are abundant and below is a high level summary [42–44].

Aplaviroc

Originally disclosed by Ono Pharmaceuticals and later developed by GSK, Aplaviroc (AK602/ONO4128/GW873140) compound **20** originated from a class of spirodiketopiperazine derivatives exemplified by compound **19** (Fig. 7). Compound 19 blocked the binding of MIP-1α to CCR5-CHO cells (IC_{50} = 2 nM) and MIP-1α

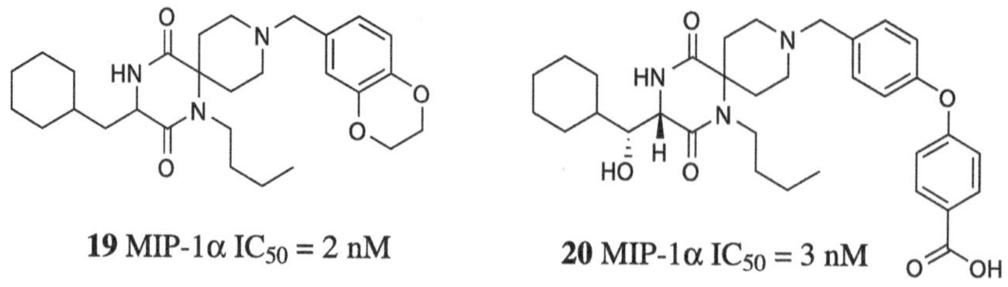

19 MIP-1α IC_{50} = 2 nM

20 MIP-1α IC_{50} = 3 nM

Figure 7

induced calcium mobilization (IC_{50} = 20 nM). Compound **19** also demonstrated potent activity against a number of R5 HIV-1 strains and multi-drug resistant strains. Compound **19** is selective for CCR5 but only modest levels of bioavailability were reported [45]. Further development resulted in Aplaviroc **20** with favorable oral bioavailability and with potency against a wide range of R5-HIV strains (IC_{50} range = 0.2–0.6 nM). Co-receptors using the CXCR4 X4-tropic HIV strains were unaffected by Aplaviroc **20** thus confirming CCR5 co-receptor selectivity. Interestingly, Aplaviroc **20** is a non-competitive allosteric antagonist that preserves RANTES and MIP-1β interactions at concentrations relevant for antiviral activity [42]. Aplaviroc **20** is a substrate for CYP3A4 and it is reported that the plasma concentrations are significantly increased when dosed in the presence of lopinavir and ritonavir [46]. All the preclinical data suggested Aplaviroc **20** as a potential therapeutic for individuals infected with HIV-1.

In Phase I/II monotherapy studies, Aplaviroc 20 decreased the viral load 1.6 $\log_{10}$ at 600 mg b.i.d. Smaller doses, i.e., 400 mg q.d. or 200 mg b.i.d. demonstrated smaller reductions [47]. Aplaviroc **20** is well tolerated up to 800 mg b.i.d. and only transient abdominal pain and nausea are reported [48]. Despite the promising early clinical results, the observation of severe liver toxicity (evidenced by elevated ALT levels) in treatment-naïve Phase IIb and treatment-experienced Phase III subjects led to the discontinuation of any further development of Aplaviroc **20** [49].

Vicriviroc

Schering-Plough's CCR5 antagonist development program produced the first compound to reach human efficacy trials with Ancriviroc (SCH-C; SC351125-compound **21**, Fig. 8). The SAR optimization toward compound **21** began with screening of in house collections which resulted in the identification of compound **4** (Fig. 1) originally part of Schering's muscarinic effort [50]. Compound **4** had mod-

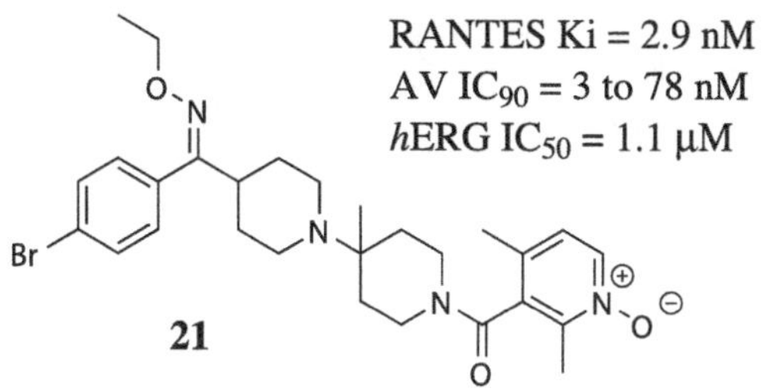

Figure 8

est CCR5 activity (RANTES K_i = 64 nM) and activity at the M_2 receptor (K_i = 230 nM). Optimization of the desired CCR5 activity and tuning out the M_2 activity resulted in compound **21** [50]. SCH-C **21** was generally well tolerated in Phase 1 safety trials although dose dependent QT_c prolongation was observed at the highest dose (600 mg). Compound **21** was shown to reduce viral load (VL) by 1.5 log_{10} in HIV infected patients at a dose of 100 mg b.i.d for 10 days [43]. Based on the affinity of compound **21** for *h*ERG and a need for improved pharmaokinetics, Schering continued in parallel to look for alternatives to compound **21** with improved antiviral and selectivity properties.

A second active compound **22** (Fig. 9), containing a piperizine scaffold was also identified during screening efforts at Schering. Compound **22** was a potent M_2 antagonist (K_i = 0.8 nM) with very modest CCR5 activity (RANTES K_i = 440 nM). Deconstructive SAR on the left-hand side of compound **22** and the observation that the (*S*)-methylpiperizine gave compounds with affinity for the CCR5 receptor with very little muscarinic activity resulted in compound **23** (RANTES K_i = 2 nM; M_2 K_i = 2500 nM). The pyridyl *N*-oxide of compound **23** has superior oral bioavailability over compound 23 but the desired selectivity for M_2 was lost (RANTES K_i = 2 nM; M_2 K_i = 250 nM) [51]. Continuing with their SAR optimization researches at Schering discovered that 4,6-dimethylpyrimidine carboxamide can replace the pyridyl N-oxide of compound **23** and introducing a benzylic methoxymethyl side chain gave the potent CCR5 antagonist SCH-D (Vicriviroc), compound **24** (RANTES K_i = 1.6 nM, Fig. 9). Selectivity over muscarinic was excellent (K_i > 10,000 nM) and *h*ERG activity was greatly reduced (IC_{50} = 5.8 μM) [52].

In Phase I studies with treatment-naïve HIV subjects, Vicriviroc **24** demonstrated dose dependent reduction in VL at 10, 25, and 50 mg b.i.d., with reductions of 1.6 log_{10} at the 50 mg dose [53]. Based on these early promising results, a Phase II study with treatment-naïve patients was initiated and later terminated when viral

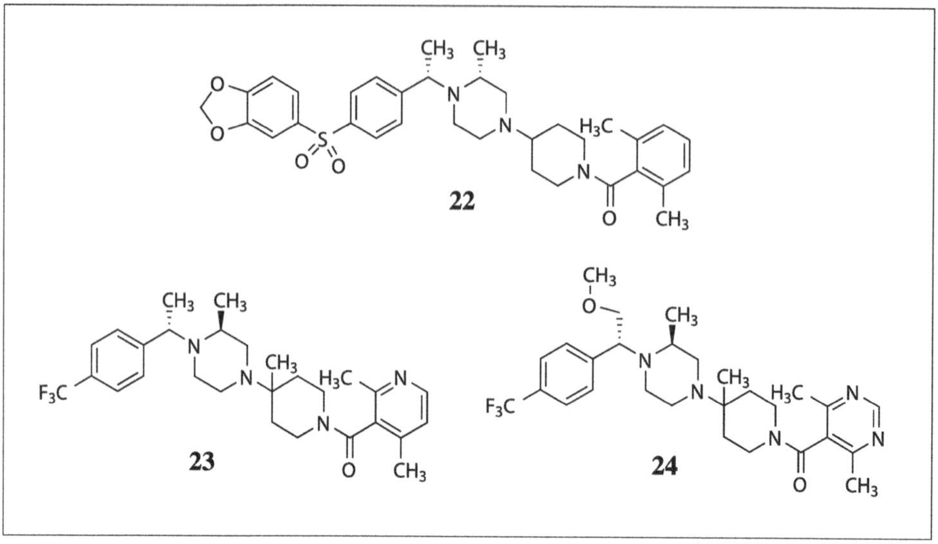

Figure 9

breakthrough was observed in the arm with Vicriviroc **24** compared to an arm with efavirenz [54]. These results suggest Vicriviroc **24** is not as effective in treatment-naïve patients *versus* current treatment options. Despite these results, a Phase II study with treatment experienced patients where Vicriviroc **24** was dosed at 5, 10 and 15 mg daily (QD) for 14 days was completed. Viral load reductions up to 1.6 $\log_{10}$ were observed at 15 mg QD, however the study was discontinued at the 5 mg dose due to lack of efficacy. The study was also un-blinded when the DSMD reported five instances of cancer among the trail participants. The decision to continue the study was made based on the strong antiviral activity and the lack of causal association of the malignancies and Vicriviroc **24** at 10 and 15 mg. Schering has initiated a new Phase II trial (VICTOR-E1) evaluating the safety and efficacy of Vicriviroc **24** at 20 and 30 mg once daily doses in treatment experienced HIV patients to determine if improved viral suppression is observed at higher doses [54, 55].

Maraviroc

Not wanting to rely on synthetic peptides or antibodies in their search for a CCR5 antagonist, researchers at Pfizer turned to high-throughput screening to identify small molecule starting points. They focused their efforts on hits with reasonable affinity and ligand efficiency [56]. The screening effort resulted in compound **6** (Fig. 1) which demonstrated weak affinity and no antiviral activity but was consid-

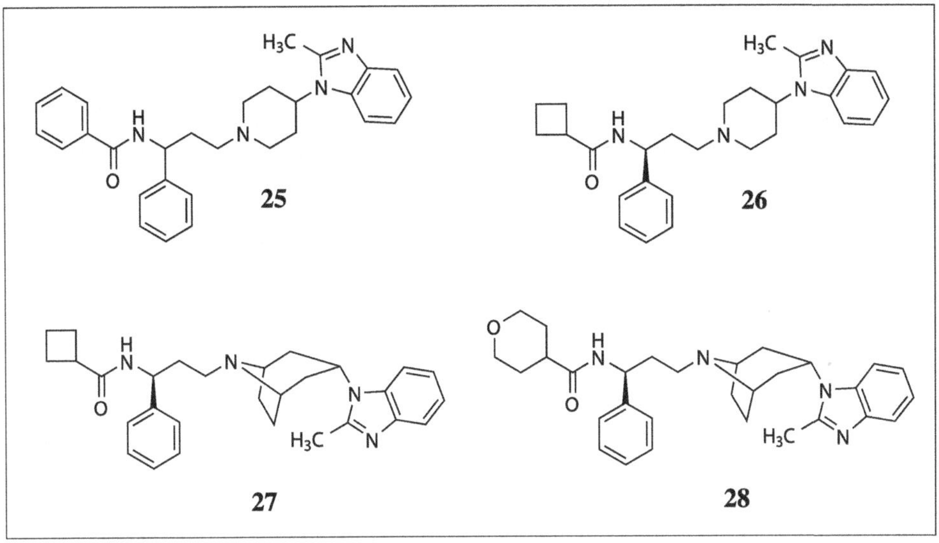

Figure 10

ered a viable starting point for further optimization [44]. Pfizer's researchers first focused on mitigating CYP2D6 activity in the original hit and reducing the lipophilicity. These efforts resulted in the replacement of the imidazopyridine with benzimidazole and swapping out the benzyhydryl group with a benzamide to give compound **25** (Fig. 10). Compound **25** demonstrated good binding potency (MIP-1β IC_{50} = 45 nM) and the beginning of antiviral activity (IC_{50} = 210 nM). SAR optimization of the amide region and identifying the enantiomeric preference of the receptor led to the cyclobutyl amide compound **26** (MIP-1β IC_{50} = 20 nM; AV IC_{50} = 73 nM). The CYP2D6 activity of compound **26** was still unacceptable and further SAR optimization determined that the [3.2.1].-azabicycloamine (tropane) can replace the aminopiperidine moiety. This resulted in compound **27** which has no activity at CYP2D6 while maintaining excellent binding affinity and antiviral activity (MIP-1β IC_{50} = 6 nM; AV IC_{90} = 3 nM). Despite these promising early results compound **27** was found to demonstrate 99% inhibition (at 1 μM) on the *h*ERG potassium channel, an activity level that is not acceptable due to the possibility of QT_c interval prolongation [57]. The Pfizer team then turned their attention to optimizing the molecule with respect to polar surface area (PSA) in an effort to dial out the *h*ERG activity while maintaining acceptable levels of cell permeability for oral absorption. The effort resulted in compound **28** which maintained desired levels of antiviral activity (AV IC_{90} = 0.6 nM) and was selective against the *h*ERG inhibition. In addition, compound **28** inhibits the binding of the three major chemokine ligands for CCR5 while not interfering with other chemokine/chemokine receptor axes.

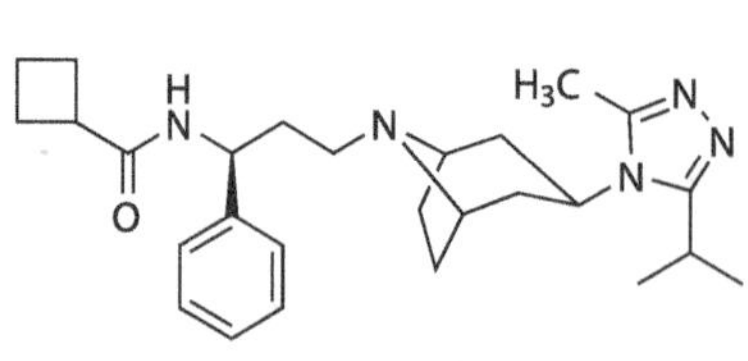

29

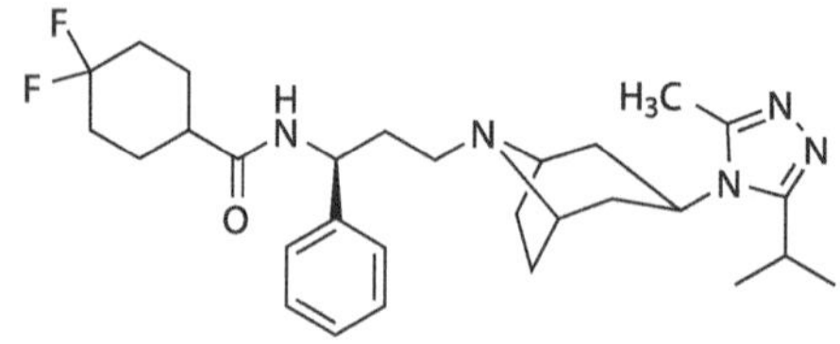

30 Maraviroc

Figure 11

Compound 28 was not cytotoxic or positive in the AMES assay, however, compound 28 demonstrated less than optimal PK properties with an oral bioavailability less than 10% in the dog.

The final push to a CCR5 antagonist with acceptable properties for further clinical development focused on reducing lipophilicity. Replacing the benzimidazole of compound 28 with a substituted triazole gave compound 29 (Fig. 11) allowing for a significant reduction in lipophilicity while maintaining the antiviral activity. Conformational constraints introduced by the isopropyl group on the triazole cause a switch in preference from the endo isomer in earlier analogs to the now favored exo isomer of compound 29. Compound 29 demonstrates acceptable PK properties in the dog with oral bioavailability at 43%. However, QT_c prolongation was once again a potential issue for compound 29 in the dog. Further work with compound 29 as the starting point demonstrated modifications to the amide region could limit metabolic liabilities and added a steric component that limited *h*ERG activity. The results of these modifications ultimately led to the 4,4'-difluorocyclohexylamide compound 30 (UK-427,857) also known as Maraviroc which demonstrated excellent antiviral potency (AV IC_{90} = 1 nM) with no *h*ERG activity when tested at 1 μM. Maraviroc 30 is active against all R5-tropic HIV strains and demonstrates excellent selectivity against other chemokine receptors and pharmacologically relevant targets [44, 58].

Maraviroc 30 demonstrated good tolerability in Phase I/II healthy volunteer clinical trials at doses from 300 to 1,200 mg daily. The subsequent pharmacokinetic studies suggested a long elimination half life (13 h) which opens the door for daily or twice daily dosing. Maraviroc 30 does not induce the CYP450 system however, co-administration with ritonavir results in a two-fold increase in exposure. In Phase IIa monotherapy studies, 82 HIV patients were given a 10 day regiment at 25, 100, or 300 mg q.d or 50, 100, or 300 mg b.i.d. A dose dependent viral load reduction was observed at all doses with maximum response in patients receiving the 300 mg b.i.d. dose with viral reduction ranging from 1.49 to 2.42 $\log_{10}$ [58].

Maraviroc **30** is in the final stages of clinical development with fully enrolled Phase III studies ongoing in both treatment naïve and treatment experienced patients with R5-tropic virus. Maraviroc **30** received an FDA Fast Track designation in July of 2005 and in contrast to Vicriviroc **24** no cancers have been noted. The DSMB has recommended that three current Phase III trials with R5-tropic virus continue as designed.

Conclusion

The late stage human efficacy trials with Maraviroc, Aplaviroc and Vicriviroc represent a significant milestone in the chemokine field as these are the first chemokine receptor antagonists to reach such a level of clinical development. Despite the setbacks with Apalviroc and Vicriviroc, the important role of CCR5 in a number of disease states warrants further pursuit of safe and efficacious CCR5 antagonists. The results from ongoing AIDS clinical trials and planned trials in immune mediated diseases will ultimately reveal the therapeutic utility of CCR5 antagonists.

References

1 Gao Z, Metz W (2003) Unraveling the chemistry of chemokine receptor ligands. *Chem Rev* 103(9): 3733–3752

2 Pease J, Williams T (2006) The attraction of chemokines as a target for specific anti-inflammatory therapy. *Brit J Pharmacol* 147 (Suppl 1): S212–S221

3 Wells T, Power C, Shaw J, Proudfoot A (2006) Chemokine blockers – therapeutics in the making? *Trends in Pharmacol Sci* 27(1): 41–47

4 Mueller A, Strange PG (2004) The chemokine receptor, CCR5. *Intl J Biochem & Cell Biol* 36(1): 35–38

5 Lederman MM, Penn-Nicholson A, Cho M, Mosier D (2006) Biology of CCR5 and its role in HIV infection and treatment. *JAMA* 296(7): 815–826

6 a) Wise A, Gearing K, Rees S (2002) Target validation of G-protein coupled receptors. *Drug Discov Today* 7(4): 235–246
b) Hopkins AL, Groom CR (2002) Opinion: The druggable genome. *Nat Rev Drug Discov* 1(9): 727–730

7 Ribeiro S, Horuk R (2005) The clinical potential of chemokine receptor antagonists. *Pharmacol & Therapeutics* 107(1): 44–58

8 Kazmierski W, Bifulco N, Yang H, Boone L, DeAnda F, Watson C, Kenakin T (2003) Recent progress in discovery of small-molecule CCR5 chemokine receptor ligands as HIV-1 inhibitors. *Bioorg & Med Chem* 11(13): 2663–2676

9 Kazmierski WM, Peckham JP, Duan M, Kenakin TP, Jenkinson S, Gudmundsson KS, Piscitelli SC, Feldman PL (2005) Recent progress in the discovery of new CCR5 and

CXCR4 chemokine receptor antagonists as inhibitors of HIV-1 entry. Part 2. *Curr Med Chem: Anti-Infective Agents* 4(2): 133–152

10 Kazmierski WM, Kenakin TP, Gudmundsson KS (2006) Peptide, peptidomimetic and small-molecule drug discovery targeting HIV-1 host-cell attachment and entry through gp120, gp41, CCR5 and CXCR4. *Chem Bio & Drug Design* 67(1): 13–26

11 Reeves JD, Piefer AJ (2005) Emerging drug targets for antiretroviral therapy. *Drugs* 65(13): 1747–1766

12 Castagna A, Biswas P, Beretta A, Lazzarin A (2005) The appealing story of HIV entry inhibitors: from discovery of biological mechanisms to drug development. *Drugs* 65(7): 879–904

13 Liu R, Paxton WA, Choe S, Ceradini D, Martin SR, Horuk R, MacDonald ME, Stuhlmann H, Koup RA, Landau NR (1996) Homozygous defect in HIV-1 coreceptor accounts for resistance of some multiply-exposed individuals to HIV-1 infection. *Cell* 86(3): 367–377

14 Connor RI, Sheridan KE, Ceradini D, Choe S, Landau NR (1997) Change in coreceptor use correlates with disease progression in HIV-1-infected individuals. *J Exp Med* 185(4): 621–628

15 Mack M, Luckow B, Nelson PJ, Cihak J, Simmons G, Clapham PR, Signoret N, Marsh M, Stangassinger M, Borlat F et al (1998) Aminooxypentane-RANTES induces CCR5 internalization but inhibits recycling: a novel inhibitory mechanism of HIV infectivity. *J Exp Med* 187(8): 1215–1224

16 Kawamura T, Bruce SE, Abraha A, Sugaya M, Hartley O, Offord RE, Arts EJ, Zimmerman PA, Blauvelt A (2004) PSC-RANTES blocks R5 human immunodeficiency virus infection of langerhans cells isolated from individuals with a variety of CCR5 diplotypes. *J Virol* 78(14): 7602–7609

17 Trkola A, Ketas TJ, Nagashima KA, Zhao L, Cilliers T, Morris L, Moore JP, Maddon PJ, Olson WC (2001) Potent, broad-spectrum inhibition of human immunodeficiency virus type 1 by the CCR5 monoclonal antibody PRO 140. *J Virol* 75(2): 579–588

18 a) Roschke V, Clark S, Branco L, Kanakaraj P, Kaufman T, Yao X, Nardelli B, Shi Y, Cai W, Ullrich A et al (2004) Characterization of a panel of novel human monoclonal antibodies that specifically antagonize CCR5 and block HIV-1 entry. 44th Annual Interscience Conference on Antimicrobial Agents and Chemotherapy (ICAAC) Abstract #2871

b) Lalezari J, Lederman PM, Dejesus E, Searle J, Cai W, Roschke V, Zhong J, Hicks C, Freimuth W, Subramanian M (2006) A phase 1, dose-escalation, placebo-controlled study of a fully human monoclonal antibody (CCR5mAb004) against CCR5 in patients with CCR5-tropic HIV-1 infection. Forty-Sixth Interscience Conference on Antimicrobial Agents and Chemotherapy, San Francisco, abstract H-1668

19 Watson C, Jenkinson S, Kazmierski W, Kenakin T (2005) The CCR5 receptor based mechanism of action of 873140, a potent allosteric noncompetitive HIV entry inhibitor. *Mol Pharmacol* 67: 1268–1282

20 a) Princen K, Hatse S, Vermeire K, Aquaro S, De Clercq E, Gerlach LO, Rosenkilde M,

Schwartz TW, Skerlj R, Bridger G et al (2004) Inhibition of human immunodeficiency virus replication by a dual CCR5 /CXCR4 antagonist. *J Virol* 78(23): 12996–13006
b) Zhou Y, Bridger GJ, Skerlj RT, Bogucki D, Yang W, Bourque E, Langille J, Li TS, Metz M (2005) Preparation of acylaminoalkylpiperidinamines as CCR5 chemokine receptor ligands. US Pat Appl Publ US 2005277668 A1

21 Palani A, Tagat JR (2006) Discovery and development of small-molecule chemokine coreceptor CCR5 antagonists. *J Med Chem* 49(10): 2851–2857

22 a) Bondensgaard K, Ankersen M, Thogersen H, Hansen BS, Wulff BS, Bywater RP (2004) Recognition of privileged structures by G-protein coupled receptors. *J Med Chem* 47(4): 888–899
b) von Korff M, Steger M (2004) GPCR-tailored pharmacophore pattern recognition of small molecular ligands. *J Chem Inf Comp Sci* 44(3): 1137–1147

23 Burrows JN, Cumming JG, Fillery SM, Hamlin GA, Hudson JA, Jackson RJ, McLaughlin S, Shaw JS (2005) Modulators of the human CCR5 receptor. Part 1: Discovery and initial SAR of 1-(3,3-diphenylpropyl)-piperidinyl amides and ureas. *Bioorg Med Chem Lett* 15(1): 25–28

24 Cumming JG, Cooper AE, Grime K, Logan CJ, McLaughlin S, Oldfield J, Shaw JS, Tucker H, Winter J, Whittaker D (2005) Modulators of the human CCR5 receptor. Part 2: SAR of substituted 1-(3,3-diphenylpropyl)-piperidinyl phenylacetamides. *Bioorg Med Chem Lett* 15(22): 5012–5015

25 Thoma G, Nuninger F, Schaefer M, Akyel KG, Albert R, Beerli C, Bruns C, Francotte E, Luyten M, MacKenzie D et al (2004) Orally bioavailable competitive CCR5 antagonists. *J Med Chem* 47(8): 1939–1955

26 Dorn CP, Finke PE, Oates B, Budhu RJ, Mills SG, MacCoss M, Malkowitz L, Springer MS, Daugherty BL, Gould SL et al (2001) Antagonists of the human CCR5 receptor as anti-HIV-1 agents. Part 1: Discovery and initial structure-activity relationships for 1-amino-2-phenyl-4-(piperidin-1-yl)butanes. *Bioorg Med Chem Lett* 11(2): 259–264

27 Mills SG, DeMartino JA (2004) Chemokine receptor-directed agents as novel anti-HIV-1 therapies. *Curr Topics Med Chem* 4(10): 1017–1033

28 Finke PE, Meurer LC, Oates B, Shah SK, Loebach JL, Mills SG, MacCoss M, Castonguay L, Malkowitz L, Springer MS et al (2001) Antagonists of the human CCR5 receptor as anti-HIV-1 agents. Part 3: A proposed pharmacophore model for 1-[N-(methyl)-N-(phenylsulfonyl)amino].-2-(phenyl)-4-[4-(substituted)piperidin-1-yl].butanes. *Bioorg Med Chem Lett* 11(18): 2469–2473
b) Castonguay LA, Weng Y, Adolfsen W, Di Salvo J, Kilburn R, Caldwell CG, Daugherty BL, Finke PE, Hale JJ, Lynch CL et al (2003) Binding of 2-aryl-4-(piperidin-1-yl)butanamines and 1,3,4-trisubstituted pyrrolidines to human CCR5: A molecular modeling-guided mutagenesis study of the binding pocket. *Biochemistry* 42(6): 1544–1550

29 a) Shen DM, Shu M, Mills SG, Chapman KT, Malkowitz L, Springer MS, Gould SL, DeMartino JA, Siciliano SJ, Kwei GY et al (2004) Antagonists of human CCR5 recep-

tor containing 4-(pyrazolyl)piperidine side chains. Part 1: Discovery and SAR study of 4-pyrazolylpiperidine side chains. *Bioorg Med Chem Lett* 14(4): 935–939

b) Shen DM, Shu M, Willoughby CA, Shah S, Lynch CL, Hale JJ, Mills SG, Chapman KT, Malkowitz L, Springer MS et al (2004) Antagonists of human CCR5 receptor containing 4-(pyrazolyl)piperidine side chains. Part 2: Discovery of potent, selective, and orally bioavailable compounds. *Bioorg Med Chem Lett* 14(4): 941–945

c) Shu M, Loebach JL, Parker KA, Mills SG, Chapman KT, Shen DM, Malkowitz L, Springer MS, Gould SL, DeMartino JA et al (2004) Antagonists of human CCR5 receptor containing 4-(pyrazolyl)piperidine side chains. Part 3: SAR studies on the benzylpyrazole segment. *Bioorg Med Chem Lett* 14(4): 947–952

30 Finke PE, Loebach JL, Parker KA, Plummer CW, Mills SG (2005) Preparation of amino acid N-cyclopentyl modulators of chemokine receptor activity. US Pat Appl Publ US 2005070609 A1

31 Veazey RS, Klasse PJ, Schader SM, Hu Q, Ketas TJ, Lu M, Marx PA, Dufour J Colonno RJ, Shattock RJ et al (2005) Protection of macaques from vaginal SHIV challenge by vaginally delivered inhibitors of virus-cell fusion. *Nature* 438(7064): 99–102

32 DeMartino JA (2006) CCR5 Blockade modulates alloimmunity in primates. Abstracts of Papers, 231st ACS National Meeting, Atlanta, GA, US, March 26–30, MEDI-193

33 a) Baba M, Nishimura O, Kanzaki N, Okamoto M, Sawada H, Iizawa Y, Shiraishi M, Aramaki Y, Okonogi K Ogawa Y et al (2000) A small-molecule, nonpeptide CCR5 antagonist with highly potent and selective anti-HIV-1 activity. *Proc Nat Acad Sci USA* 96(10): 5698–5703

b) Shiraishi M, Aramaki Y, Seto M, Imoto H, Nishikawa Y, Kanzaki N, Okamoto M, Sawada H, Nishimura O, Baba M et al (2000) Discovery of novel, potent, and selective small-molecule CCR5 antagonists as anti-HIV-1 agents: synthesis and biological evaluation of anilide derivatives with a quaternary ammonium moiety. *J Med Chem* 43(10): 2049–2063

c) Kanzaki N, Shiraishi M, Lizawa Y, Baba M, Nishimura O, Fujino M (2000) TAK-779 anti-HIV chemokine CCR5 antagonist. *Drugs of the Future* 25(3): 252–258

34 Baba M, Takashima K, Miyake H, Kanzaki N, Teshima K, Wang X, Shiraishi M, Iizawa Y (2005) TAK-652 inhibits CCR5-mediated human immunodeficiency virus type 1 infection *in vitro* and has favorable pharmacokinetics in humans. *Antimicrob Agents Chemother* 49(11): 4584–4591

35 Takashima K, Miyake H, Kanzaki N, Tagawa Y, Wang X, Sugihara Y, Iizawa Y, Baba M (2005) Highly potent inhibition of human immunodeficiency virus type 1 replication by TAK-220, an orally bioavailable small-molecule CCR5 antagonist. *Antimicrob Agents Chemother* 49(8): 3474–3482

36 Nishikawa M, Takashima K, Nishi T, Furuta RA, Kanzaki N, Yamamoto Y, Fujisawa J (2005) Analysis of binding sites for the new small-molecule CCR5 antagonist TAK-220 on human CCR5. *Antimicrob Agents Chemother* 49(11): 4708–4715

37 Tremblay CL, Giguel F, Guan Y, Chou TC, Takashima K, Hirsch MS (2005) TAK-220, a novel small-molecule CCR5 antagonist, has favorable anti-human immunodeficiency

virus interactions with other antiretrovirals *in vitro*. *Antimicrob Agents Chemother* 49(8): 3483–3485

38 a) Kazmierski WM, Aquino CJ, Bifulco N, Boros EE, Chauder BA, Chong PY, Duan M, Deanda F, Koble CS, Mclean EW et al (2004) Prepn of benzimidazolyl azabicyclooctyl ethylpiperidines as CCR5 antagonists for the treatment of HIV infection. PCT Int Appl WO 2004054974 A2
b) Duan M, Kazmierski WM, Aquino CJ (2004) Preparation of heterocyclylalkyl substituted cyclohexyl compounds as CCR5 antagonists. PCT Int Appl WO 2004054581 A2

39 Kazmierski WM, Anderson DL, Aquino CJ, Bifulco N, Boone LR, Boros EE, Chauder BA, Chong PY, Duan M, Ferris RG et al (2006) Novel, potent and bioavailable CCR5 chemokine receptor small-molecule antagonists for HIV therapy: Scaffold discovery and addressing hERG ion channel affinity in the process of optimizing potency and bioavailability. Abstracts of Papers, 232nd ACS National Meeting, San Francisco, CA, US, Sept. 10–14, 2006 MEDI-297

40 a) Xue CB, Cao G, Huang T, Chen L, Zhang K, Wang A, Meloni D, Anand R, Glenn J, Metcalf BW (2005) Preparation of piperazinylpiperidine derivatives as chemokine receptor antagonists. US Pat Appl Publ US 2005261310 A1
b) IDdb (2006) INCB-9471. The Investigational Drugs Database Drug report

41 http://www.incyte.com/backgrounders%208.10.06/CCR5%20Backgrounder.sept.pdf

42 Maeda K, Nakata H, Koh Y, Miyakawa T, Ogata H, Takaoka Y, Shibayama S, Sagawa K, Fukushima D, Moravek J et al (2004) Spirodiketopiperazine-based CCR5 inhibitor which preserves CC-chemokine/CCR5 interactions and exerts potent activity against R5 human immunodeficiency virus type 1 *in vitro*. *J Virol* 78(16): 8654–8662

43 Strizki JM, Tremblay C, Xu S, Wojcik L, Wagner N, Gonsiorek W, Hipkin RW, Chou CC, Pugliese-Sivo C, Xiao Y et al (2005). Discovery and characterization of vicriviroc (SCH 417690), a CCR5 antagonist with potent activity against human immunodeficiency virus type 1. *Antimicrob Agents Chemother* 49(12): 4911–4919

44 a) Wood A, Armour D (2005) The discovery of the CCR5 receptor antagonist, UK-427,857, a new agent for the treatment of HIV infection and AIDS. *Prog Med Chem* 43: 239–271
b) Dorr P, Westby M, Dobbs S, Griffin P, Irvine B, Macartney M, Mori J, Rickett G, Smith-Burchnell C, Napier C et al (2005) Maraviroc (UK-427,857), a potent, orally bioavailable, and selective small-molecule inhibitor of chemokine receptor CCR5 with broad-spectrum anti-human immunodeficiency virus type 1 activity. *Antimicrob Agents Chemother* 49(11): 4721–4732

45 Maeda K, Yoshimura K, Shibayama S, Habashita H, Tada H, Sagawa K, Miyakawa T, Aoki M, Fukushima D, Mitsuya H (2001) Novel low molecular weight spirodiketopiperazine derivatives potently inhibit R5 HIV-1 infection through their antagonistic effects on CCR5. *J Biol Chem* 276(37): 35194–35200

46 Adkison KK, Shachoy-Clark A, Fang L, Lou Y, Otto VR, Berrey MM, Piscitelli SC (2006) The effects of ritonavir and lopinavir/ritonavir on the pharmacokinetics of a

novel CCR5 antagonist, aplaviroc, in healthy subjects. *British J Clinical Pharmacol* 62(3): 336–344

47 Lalezari J, Thompson M, Kumar P, Piliero P, Davey R, Patterson K, Shachoy-Clark A, Adkison K, Demarest J, Lou Y et al (2005) Antiviral activity and safety of 873140, a novel CCR5 antagonist, during short-term monotherapy in HIV-infected adults. *AIDS* 19(14): 1443–1448

48 Adkison KK, Shachoy-Clark A, Fang L, Lou Y, O'Mara K, Berrey MM, Piscitelli SC (2005) Pharmacokinetics and short-term safety of 873140, a novel CCR5 antagonist, in healthy adult subjects. *Antimicrob Agents Chemother* 49(7): 2802–2806

49 a) GlaxoSmithKline (2005) Statement to HIV patient community: information from GlaxoSmithKline on changes to studies of investigational CCR5 entry inhibitor Aplaviroc (GW873140). 15 September 2005

b) IDdb (2006) Aplaviroc. The Investigational Drugs Database Drug reports

50 a) Palani A, Shapiro S, Clader JW, Greenlee WJ, Cox K, Strizki J, Endres M, Baroudy BM (2001) Discovery of 4-[(Z)-(4-Bromophenyl)(ethoxyimino)methyl].-1'-[(2,4-dimethyl-3-pyridinyl)carbonyl].-4'-methyl-1,4'-bipiperidine N-Oxide (SCH 351125): An orally bioavailable human CCR5 antagonist for the treatment of HIV infection. *J Med Chem* 44(21): 3339–3342

b) Strizki JM, Xu S, Wagner NE, Wojcik L, Liu J, Hou Y, Endres M, Palani A, Shapiro S, Clader JW et al (2001) SCH-C (SCH 351125), an orally bioavailable, small molecule antagonist of the chemokine receptor CCR5, is a potent inhibitor of HIV-1 infection *in vitro* and *in vivo*. *Proc Nat Acad Sci* 98(22): 12718–12723

51 a) Tagat JR, McCombie SW, Steensma RW, Lin SI, Nazareno DV, Baroudy B, Vantuno N, Xu S, Liu J (2001) Piperazine-based CCR5 antagonists as HIV-1 inhibitors. I: 2(S)-methyl piperazine as a key pharmacophore element. *Bioorg Med Chem Lett* 11(16): 2143–2146

b) Tagat JR, Steensma RW, McCombie SW, Nazareno DV, Lin SI, Neustadt BR, Cox K, Xu S, Wojcik L, Murray MG et al (2001) Piperazine-based CCR5 antagonists as HIV-1 inhibitors. II. discovery of 1-[(2,4-dimethyl-3-pyridinyl)carbonyl].-4-methyl-4-[3(S)-methyl-4-[1(S)-[4-(trifluoro-methyl)phenyl].ethyl].-1-piperazinyl].-piperidine N1-oxide (Sch-350634), an orally bioavailable, potent CCR5 antagonist. *J Med Chem* 44(21): 3343–3346

52 a) McCombie SW, Tagat JR, Vice SF, Lin SI, Steensma R, Palani A, Neustadt BR, Baroudy BM, Strizki JM, Endres M et al (2003) Piperazine-based CCR5 antagonists as HIV-1 inhibitors. III: synthesis, antiviral and pharmacokinetic profiles of symmetrical heteroaryl carboxamides. *Bioorg Med Chem Lett* 13(3): 567–571

b) Tagat JR, McCombie SW, Nazareno D, Labroli MA, Xiao Y, Steensma RW, Strizki JM, Baroudy BM, Cox K, Lachowicz J et al (2004) Piperazine-based CCR5 antagonists as HIV-1 inhibitors. IV. Discovery of 1-[(4,6-dimethyl-5-pyrimidinyl)carbonyl].-4-[4-{2-methoxy-1(R)-4-(trifluoromethyl)phenyl}ethyl-3(S)-methyl-1-piperazinyl].-4-methylpiperidine (Sch-417690/Sch-D), a potent, highly selective, and orally bioavailable CCR5 antagonist. *J Med Chem* 47(10): 2405–2408

53 a) Schurmann D et al (2004) SCH D: antiviral activity of a CCR5 receptor antagonist. *Eleventh Conference on Retroviruses and Opportunistic Infections*, San Francisco, abstract 140LB
b) http://www.aidsmap.com/en/docs/8894CDC2-2CDD-4C48-A510-678FBB89EE62.asp
54 a) Thomson-Pharma[SM] Drug Report; 2006 Vicriviroc. https://www.thomson-pharma.com/portal/page
b) Greaves W, Landovitz R, Fatkenheuer G et al (2004) Late virologic breakthrough in treatment-naive patients on a regimen of Combivar + Vicriviroc. *Proceedings of the 11th Conference on Retroviruses and Opportunistic Infections*. San Francisco, CA, USA, 8–11 February 2004 (Abstract 161LB)
55 a) Gulick R, Su Z, Flexner C et al (2006) ACTG 5211: Phase II study of the safety and efficacy of Vicriviroc in HIV-infected treatment-experienced subjects. *16th International AIDS Conference*. 13–18 August 2006. Toronto, Canada. Abstract THLB0217
b) Schering-Plough. Vicriviroc demonstrates potent and sustained viral suppression in ACTG Phase II clinical study in treatment-experienced HIV patients. *Press Release* 17 August 2006
c) For additional information on the VICTOR-E1 trial, see http://www.clinicaltrials.gov
56 Kuntz ID, Chen K, Sharp KA, Kollman PA (1999) The maximal affinity of ligands. *Proc Natl Acad Sci* 96(18): 9997–10002
57 Fermini B, Fossa AA (2003) The impact of drug-induced QT interval prolongation on drug discovery and development. *Nat Rev Drug Discov* 2(6): 439–447
58 Overton ET, Powderly WG (2006) Future of Maraviroc and other CCR5 antagonists. *Future Virology* 1(5): 605–613

Index

The PIR-Series
Progress in Inflammation Research

Homepage: http://www.birkhauser.ch

Up-to-date information on the latest developments in the pathology, mechanisms and therapy of inflammatory disease are provided in this monograph series. Areas covered include vascular responses, skin inflammation, pain, neuroinflammation, arthritis cartilage and bone, airways inflammation and asthma, allergy, cytokines and inflammatory mediators, cell signalling, and recent advances in drug therapy. Each volume is edited by acknowledged experts providing succinct overviews on specific topics intended to inform and explain. The series is of interest to academic and industrial biomedical researchers, drug development personnel and rheumatologists, allergists, pathologists, dermatologists and other clinicians requiring regular scientific updates.

Available volumes:

T Cells in Arthritis, P. Miossec, W. van den Berg, G. Firestein (Editors), 1998
Chemokines and Skin, E. Kownatzki, J. Norgauer (Editors), 1998
Medicinal Fatty Acids, J. Kremer (Editor), 1998
Inducible Enzymes in the Inflammatory Response,
D.A. Willoughby, A. Tomlinson (Editors), 1999
Cytokines in Severe Sepsis and Septic Shock, H. Redl, G. Schlag (Editors), 1999
Fatty Acids and Inflammatory Skin Diseases, J.-M. Schröder (Editor), 1999
Immunomodulatory Agents from Plants, H. Wagner (Editor), 1999
Cytokines and Pain, L. Watkins, S. Maier (Editors), 1999
In Vivo *Models of Inflammation*, D. Morgan, L. Marshall (Editors), 1999
Pain and Neurogenic Inflammation, S.D. Brain, P. Moore (Editors), 1999
Anti Inflammatory Drugs in Asthma, A.P. Sampson, M.K. Church (Editors), 1999
Novel Inhibitors of Leukotrienes, G. Folco, B. Samuelsson, R.C. Murphy (Editors), 1999
Vascular Adhesion Molecules and Inflammation, J.D. Pearson (Editor), 1999
Metalloproteinases as Targets for Anti-Inflammatory Drugs,
K.M.K. Bottomley, D. Bradshaw, J.S. Nixon (Editors), 1999
Free Radicals and Inflammation, P.G. Winyard, D.R. Blake, C.H. Evans (Editors), 1999
Gene Therapy in Inflammatory Diseases, C.H. Evans, P. Robbins (Editors), 2000
New Cytokines as Potential Drugs, S. K. Narula, R. Coffmann (Editors), 2000
High Throughput Screening for Novel Anti-inflammatories, M. Kahn (Editor), 2000
Immunology and Drug Therapy of Atopic Skin Diseases,
C.A.F. Bruijnzeel-Komen, E.F. Knol (Editors), 2000
Novel Cytokine Inhibitors, G.A. Higgs, B. Henderson (Editors), 2000
Inflammatory Processes. Molecular Mechanisms and Therapeutic Opportunities,
L.G. Letts, D.W. Morgan (Editors), 2000

Cellular Mechanisms in Airways Inflammation, C. Page, K. Banner, D. Spina (Editors), 2000
Inflammatory and Infectious Basis of Atherosclerosis, J.L. Mehta (Editor), 2001
Muscarinic Receptors in Airways Diseases, J. Zaagsma, H. Meurs, A.F. Roffel (Editors), 2001
TGF-β and Related Cytokines in Inflammation, S.N. Breit, S. Wahl (Editors), 2001
Nitric Oxide and Inflammation, D. Salvemini, T.R. Billiar, Y. Vodovotz (Editors), 2001
Neuroinflammatory Mechanisms in Alzheimer's Disease. Basic and Clinical Research, J. Rogers (Editor), 2001
Disease-modifying Therapy in Vasculitides, C.G.M. Kallenberg, J.W. Cohen Tervaert (Editors), 2001
Inflammation and Stroke, G.Z. Feuerstein (Editor), 2001
NMDA Antagonists as Potential Analgesic Drugs, D.J.S. Sirinathsinghji, R.G. Hill (Editors), 2002
Migraine: A Neuroinflammatory Disease? E.L.H. Spierings, M. Sanchez del Rio (Editors), 2002
Mechanisms and Mediators of Neuropathic pain, A.B. Malmberg, S.R. Chaplan (Editors), 2002
Bone Morphogenetic Proteins. From Laboratory to Clinical Practice, S. Vukicevic, K.T. Sampath (Editors), 2002
The Hereditary Basis of Allergic Diseases, J. Holloway, S. Holgate (Editors), 2002
Inflammation and Cardiac Diseases, G.Z. Feuerstein, P. Libby, D.L. Mann (Editors), 2003
Mind over Matter – Regulation of Peripheral Inflammation by the CNS, M. Schäfer, C. Stein (Editors), 2003
Heat Shock Proteins and Inflammation, W. van Eden (Editor), 2003
Pharmacotherapy of Gastrointestinal Inflammation, A. Guglietta (Editor), 2004
Arachidonate Remodeling and Inflammation, A.N. Fonteh, R.L. Wykle (Editors), 2004
Recent Advances in Pathophysiology of COPD, P.J. Barnes, T.T. Hansel (Editors), 2004
Cytokines and Joint Injury, W.B. van den Berg, P. Miossec (Editors), 2004
Cancer and Inflammation, D.W. Morgan, U. Forssmann, M.T. Nakada (Editors), 2004
Bone Morphogenetic Proteins: Bone Regeneration and Beyond, S. Vukicevic, K.T. Sampath (Editors), 2004
Antibiotics as Anti-Inflammatory and Immunomodulatory Agents, B.K. Rubin, J. Tamaoki (Editors), 2005
Antirheumatic Therapy: Actions and Outcomes, R.O. Day, D.E. Furst, P.L.C.M. van Riel, B. Bresnihan (Editors), 2005
Regulatory T-Cells in Inflammation, L. Taams, A.N. Akbar, M.H.M Wauben (Editors), 2005
Sodium Channels, Pain, and Analgesia, K. Coward, M. Baker (Editors), 2005
Turning up the Heat on Pain: TRPV1 Receptors in Pain and Inflammation, A.B Malmberg, K.R. Bley (Editors), 2005
The NPY Family of Peptides in Immune Disorders, Inflammation, Angiogenesis and Cancer, Z. Zukowska, G. Z. Feuerstein (Editors), 2005
Toll-like Receptors in Inflammation, L.A.J. O'Neill, E. Brint (Editors), 2005
Complement and Kidney Disease, P. F. Zipfel (Editor), 2006
Chemokine Biology – Basic Research and Clinical Application, Volume 1: Immunobiology of Chemokines, B. Moser, G. L. Letts, K. Neote (Editors), 2006
The Hereditary Basis of Rheumatic Diseases, R. Holmdahl (Editor), 2006
Lymphocyte Trafficking in Health and Disease, R. Badolato, S. Sozzani (Editors), 2006
In Vivo *Models of Inflammation, 2nd Edition, Volume I*, C.S. Stevenson, L.A. Marshall, D.W. Morgan (Editors), 2006
In Vivo *Models of Inflammation, 2nd Edition, Volume II*, C.S. Stevenson, L.A. Marshall, D.W. Morgan (Editors), 2006

GPSR Compliance

The European Union's (EU) General Product Safety Regulation (GPSR) is a set of rules that requires consumer products to be safe and our obligations to ensure this.

If you have any concerns about our products, you can contact us on ProductSafety@springernature.com

In case Publisher is established outside the EU, the EU authorized representative is:

Springer Nature Customer Service Center GmbH
Europaplatz 3
69115 Heidelberg, Germany

Batch number: 10578049

Printed by Printforce, the Netherlands